iTQ for IT Users

Using Office 2007

Jackie Sherman
Richard McGill

www.pearsonschoolsandfe.co.uk

✓ Free online support
✓ Useful weblinks
✓ 24 hour online ordering

0845 630 44 44

Heinemann

Part of Pearson

Heinemann is an imprint of Pearson Education Limited, a company incorporated in England and Wales, having its registered office at Edinburgh Gate, Harlow, Essex, CM20 2JE. Registered company number: 872828

www.heinemann.co.uk

Heinemann is a registered trademark of Pearson Education Limited

Text © Pearson Education Limited 2010

First published 2010

14 13 12 11 10
10 9 8 7 6 5 4 3 2 1

British Library Cataloguing in Publication Data
A catalogue record for this book is available from the British Library.

ISBN 978 043546871 2

Typeset by Tek-Art, Crawley Down, West Sussex
Cover design by Wooden Ark Studios
Cover photo/illustration © Photolibrary / Blend Images
Printed in the UK by Scotprint

Acknowledgements
The author and publisher would like to thank the following for their kind permission to reproduce their photographs:

Alamy Images: Art Directors & TRIP p142 (centre right); **Pearson Education Ltd:** Steve Shott p115; **Shutterstock:** Ingvar Bjork p6, iofoto p142 (bottom), Sean Gladwell p114, Vladimir Mucibabic p142 (centre left).

All other images © Pearson Education

Every effort has been made to contact the copyright holders of material reproduced in this book. Any omissions will be rectified in subsequent printings if notice is given to the publishers.

Microsoft product screenshots reprinted with permission from Microsoft Corporation. Microsoft, Encarta, MSN, and Windows are either registered trademarks or trademarks of Microsoft Corporation in the United States and/or other countries.

Thanks also to the following organizations for their permission to reprint screenshots: Sony Ericsson p7; Yahoo!/Flickr p132; Mozy p133; GMX pp135-140; Skype p141.

Whilst the publisher has taken all reasonable care in the preparation of this book, the City and Guilds of London Institute makes no representation, express or implied, with regard to the accuracy of the information contained in this book. The City and Guilds of London Institute do not accept any legal responsibility or liability for any errors or omissions from the book or the consequences thereof.

Contents

* A number of units in this book also support ISF FS IT software fundamentals.

Introduction

The qualification

ITQ is a suite of qualifications offered by City & Guilds aimed at improving skills and confidence in the general areas of Information Technology (IT) such as word processing, email, the Internet and databases.

The qualification can be taken at one of three levels, depending on your abilities and experience.

Level 1 – Award for IT Users: builds basic skills and confidence

Level 2 – Certificate for IT Users: develops existing skills and widens knowledge

Level 3 – Diploma for IT Users: combines planning and organising with a high level of IT knowledge

To achieve the **Level 1 Award**, you must gain 9 credits from any unit, a minimum of 6 to come from any Level 1 units.

To achieve the **Level 1 Certificate**, you must gain 13 credits: 3 from completing the mandatory unit, a minimum of 5 from the optional units at Level 1 or above, plus 5 credits from any optional units offered.

To achieve the **Level 1 Diploma**, you must gain 37 credits: 3 from completing the mandatory unit, a minimum of 17 from the optional units at Level 1 or above, plus 17 credits from any optional units offered.

To achieve the **Level 2 Award**, you must gain 10 credits from any unit, a minimum of 7 to come from any Level 2 units.

To achieve the **Level 2 Certificate**, you must gain 16 credits: 4 from completing the mandatory unit, a minimum of 6 from the optional units at Level 2 or above, plus 6 credits from any optional units offered.

To achieve the **Level 2 Diploma**, you must gain 38 credits: 4 from completing the mandatory unit, a minimum of 17 from the optional units at Level 2 or above, plus 17 credits from any optional units offered.

To achieve the **Level 3 Award**, you must gain 12 credits from any unit, a minimum of 8 to come from any Level 3 units.

To achieve the **Level 3 Certificate**, you must gain 25 credits: 5 from completing the mandatory unit, a minimum of 10 from the optional units at Level 3 or above, plus 10 credits from any optional units offered.

To achieve the **Level 3 Diploma**, you must gain 39 credits: 5 from completing the mandatory unit, a minimum of 17 from the optional units at Level 2 or above, plus 17 credits from any optional units offered.

The mandatory unit for ITQ has not been given separate coverage within this book. The core principals of Improving Productivity Using IT apply to all of the aspects of IT that you will learn about in order to obtain a qualification. As a result, you can achieve the mandatory unit by completing the assessments for all units and keeping a separate log of improvements you have made to your productivity over the course of the qualification.

Study normally takes place at a training centre or college, and assessments can be a mixture of online multiple choice tests and centre-assessed practical tasks.

Full details of the qualifications and accredited study centres can be found at www.cityandguilds.com.

Appropriate software

In 2007 Microsoft released a completely new version of their Office suite of programs that includes Word, Excel and PowerPoint. Until then, new versions of the programs did not differ fundamentally from the previous ones. However, with Office 2007 there has been such a complete overhaul of the look and workings of the software that it is hard for people accustomed to Office 2000, XP or 2003 to use it without difficulty. For this reason, although you can gain the ITQ qualification using any appropriate software, we have taken the opportunity to introduce readers to Office 2007 programs, as well as the latest version of Internet Explorer.

This publication therefore covers the following units:

Core Unit:

Improving Productivity Using IT

Optional Units:

IUF FS IT user fundamentals/OSP Optimise IT system performance

WP Word processing software

SS Spreadsheet software

DB Database software

INT IT communication fundamentals/INT Using the Internet

PS Presentation software

PIM Personal information management software

Using this book

For each unit, you will find full coverage of the underpinning theoretical knowledge and step-by-step guidance on the practical skills needed to carry out the everyday tasks required for the qualification.

After each section, you will be offered the chance to complete a short exercise to check your understanding, show that you have gained proficiency in the required skills and help identify any gaps in your knowledge. There is also a final assignment for each unit, designed to reflect the style and coverage of questions you will face when taking the assessments.

In many of the units, you will also find extra material. Although not required for the qualification, this will introduce you to related topics and can be used to expand your basic skills and competencies.

Throughout the book, instructions involving clicking on screen, pressing keys on the keyboard or selecting menu options will be shown in **bold**. For example, when you are asked to open the File menu and select the Save option, this will be shown as: Go to **File – Save**.

Some of the exercises in this book refer to documents on the accompanying CD-ROM.

UNIT IUF FS

IT user fundamentals/ Optimise system performance

This unit develops a more in depth understanding of the operation of computers and the operating system in everyday use. It is concerned in particular with input and output devices, the different types of computer, how to organise directories and folders, desktop configuration and health and safety issues related to using a computer.

In particular, you will be able to:

➔ prepare peripheral devices and hardware for use

➔ maintain directory/folder structures

➔ use the operating environment

➔ identify health and safety issues and good practice.

Computer configuration

The way your computer has been set up, or **configured**, will affect its function and performance, and so it is important to be able to find out these details. You will also need this information if you have a problem that you want a helpdesk to sort out, or if you need to add to your system by choosing the software or hardware that will function most effectively.

determine your computer's configuration

1 Right Click on **My Computer**.

2 Select **Properties** and click on the **General** tab.

3 This will give you very basic system information in terms of the computer's name, its processor type and speed and the amount of RAM.

4 For a more detailed view of the hardware installed on your computer, select the **Hardware** tab and click on the **Device Manager** button.

5 You will see a list of all hardware resources running on your computer and can click on a **plus sign** for more details – for example, about the modem, disk drives or sound card.

6 If any component is not working properly this will be highlighted by a yellow exclamation mark. Any disabled components will be highlighted with a red cross.

Fig. 21.1 Configuration

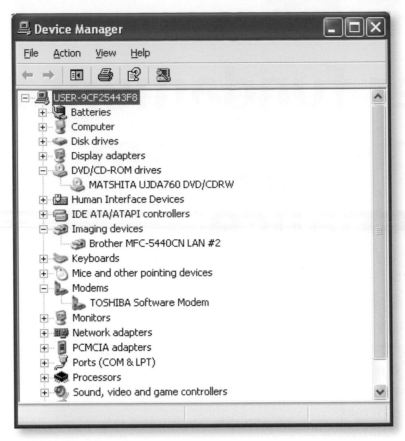

Fig. 21.2 Device manager

7 For even more detail, go to **Start – Run** and type *Winmsd* into the box. This will open a System Information window.

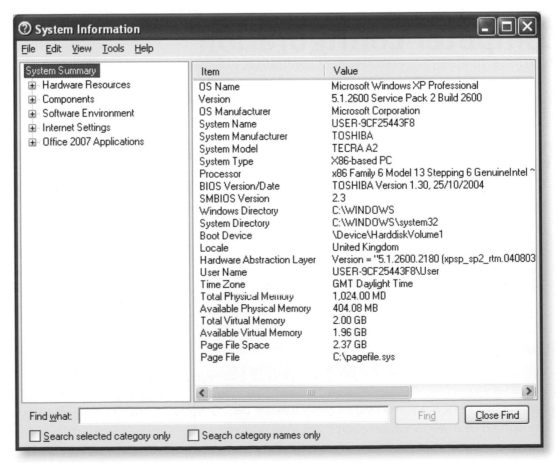

Fig. 21.3 Winmsd

8 To check the monitor settings, go to **Start – Control Panel – Display** and click on the **Settings** tab.

9 Here you can make changes to the colour quality or screen resolution – for example, from low (800 x 600 pixels) to higher (1024 x 768 pixels).

10 Click on **Advanced – Monitor** to check the refresh rate of the monitor. This is a measure of how many times a second the screen redraws data, which can be particularly important when playing some games or watching videos.

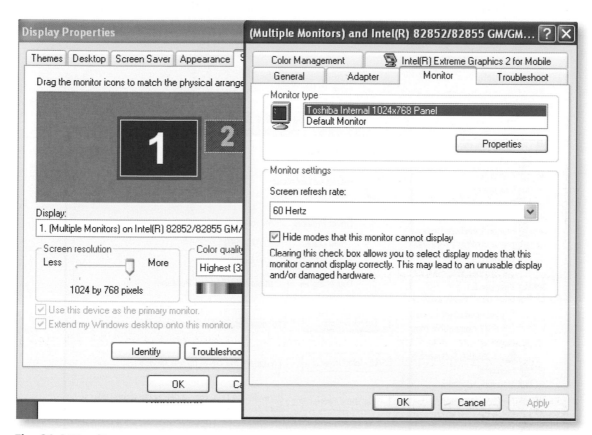

Fig. 21.4 Monitor

11 Click on tabs such as **Troubleshoot** if you want to correct particular problems with the monitor.

Monitor distortion

If the picture is distorted on a TFT monitor, it may be because you have not configured the settings to the native resolution of the screen, which is the only resolution at which the display can be correct and sharp. A 15" screen would almost always have a native resolution of 1024 x 768 pixels whereas a 17" screen would usually be 1280 x 1024 pixels.

(In the past, screensavers were used to protect monitors from burn-in when the same image was displayed for too long. This is not really a problem today but screensavers are still useful if you do not want your work on view when you leave the computer for a short while, or simply for entertainment.)

Drivers

For hardware such as the printer, speakers, VDU, mouse and keyboard to work properly, the system needs to run the correct software or driver. This is normally installed when you set up the hardware or would have been when the computer was first configured so that equipment such as the keyboard and mouse could be used.

As manufacturers regularly improve and update drivers, your original software can become out of date. If you are getting a poor performance from any hardware, it is usually suggested that you download the latest driver as this is often all that is needed to correct the problem.

view and update drivers

1   From the **Device Manager** window, right click on the hardware item and select **Properties**.

2   On the **Driver** tab you will see which driver is installed and there are buttons to let you view the details or update the driver.

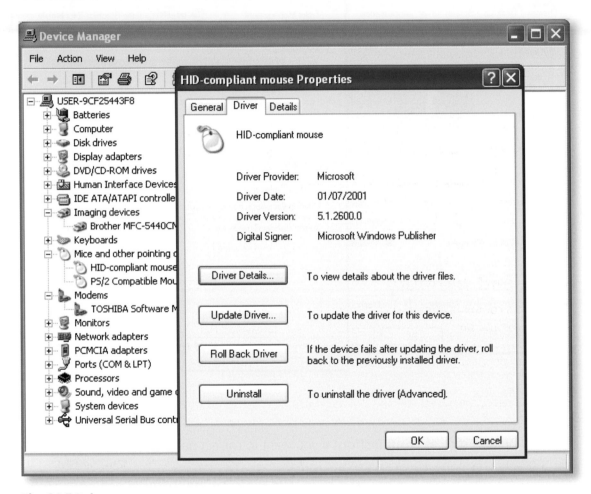

Fig. 21.5 Driver

Or

3   Select the **Update** option directly after right clicking on the hardware.

4   A wizard will take you through the steps.

Or

5   Visit the manufacturer's website and click on the link to **Drivers**.

6   Locate your particular item of hardware from the list and download the driver appropriate for your operating system.

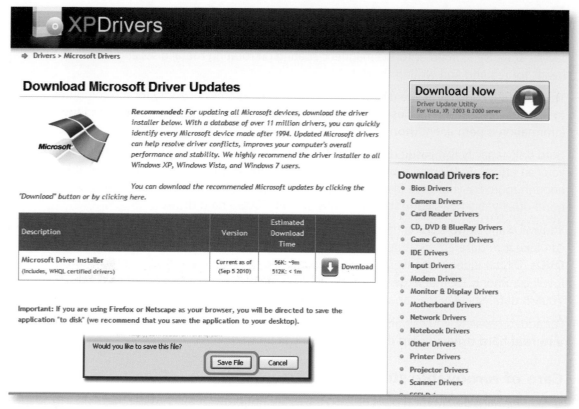

Fig. 21.6 Download driver

CPU and memory

Computers contain a printed circuit board, the motherboard, which houses various integrated circuit chips and provides expansion slots for extra hardware. The most important chip is the **microprocessor** or **CPU** (Central Processing Unit) – the 'brains' of the computer. This is where all the thinking, calculating and processing takes place. Well-known brands of CPU include Intel and AMD.

The CPU speed, measured in **megahertz** (MHz) or **gigahertz** (GHz), determines the performance of the computer, as it is a measure of how many millions of instructions per second the processor can carry out.

Other chips contain the computer's **main memory**. This is known as **RAM** (Random Access Memory) and is a **volatile** (temporary) store for data or instructions typed into the computer, as well as the programs currently in use. When you switch off the computer, the contents of RAM are lost.

Data storage

The hard disk or hard drive is a sealed unit built inside the machine that is the main storage facility. Data is recorded onto the magnetic coating on rotating metal discs and there are also heads for reading and writing the data.

The hard disk contains all the digital information and programs necessary for the computer to operate effectively. Any files you produce as you work can also be stored here, and the information is permanent (**non-volatile**) as it is retained even when the computer is switched off.

Hard disk capacity is measured in **gigabytes** (GB) and for home use you should buy the largest you can afford. Today new programs can take up hundreds of megabytes so you must have enough room for your own files. If you search the Internet for new computers, you will find that good-quality machines currently come with 500GB–750GB hard drives.

You will want to store some data on removable storage devices such as optical disks so that you can take it to different locations. The two types of disk you will use are **CDs** (Compact Disks) and **DVDs** (Digital Video/Versatile Disks) and the data they hold will vary depending on your needs. It will commonly include large work files, photos, games, music or films. CDs usually hold about 700MB data whereas the capacity of DVDs can be 4.7Gb or higher.

For added capacity, you can also use plug-in peripherals such as **flash drives** or **memory sticks, external hard drives** or **zip drives** that hold compressed data.

Care of removable storage devices

Unlike the hard disk, optical disks or flash drives are not protected inside the machine and so are very vulnerable to heat, dust, liquids or magnetism as they will be carried around and left on desks or in drawers. You could lose all your data if these devices are not handled carefully and protected from harm.

Ports and connectors

The various connectors and ports on the computer allow it to communicate with the peripherals you attach. The connectors on the back of your computer are called **input/output ports** or **communication ports** and they will receive the appropriate connectors on hardware items or cables. The most common type of connector is a **USB**, but especially in older machines you will also find serial and parallel ports – for example, for plugging in a mouse, modem or keyboard.

Installing input and output devices

As you will have learned in Level 1, a wide range of peripherals or hardware items enable you to carry out tasks on the computer. These include devices such as a keyboard, mouse, mobile phone/camera or scanner to input instructions or images into the machine, and printers, speakers or plotters to output data in the form of sounds or hard copy.

All items must be installed correctly if they are to be compatible and work effectively with your computer. New purchases such as mobile phones, cameras or scanners usually come with an installation disk so you simply open this and follow the on-screen instructions. You can also plug in new hardware – often into a USB port – and use the Windows wizard to work through installation directly.

install mobile phone software from a CD

1 Insert the disk and click on **Install**.
2 Follow the instructions to install the software, possibly including additional programs that allow you to view or read files.
3 You will need to accept the terms and conditions.

4 There is usually an option to choose a full install or to select partial installation.
5 You will be told whether or not it is safe to plug in the hardware *before* installation is complete.

Fig. 21.7 Install

6 After installation, you may need to restart your computer before you can start working with the device.

Fig. 21.8 Phone suite installed

install a printer using the wizard

1 Connect the new hardware and you will usually see a message saying that it has been detected.

2 Click on the balloon for further information and then follow the instructions. If you don't have an installation disk, you can click to go online and use an automatic search for the software.

Fig. 21.9 Found hardware

Fig. 21.10 Printer install 1

3 If it is not found automatically, you may be able to select the **Advanced** option which will allow you to select the make and model of the hardware and locate the appropriate software that way.

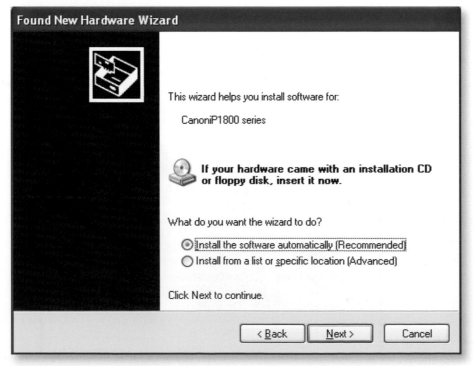

Fig. 21.11 Printer install 2

Check your understanding 1

1 Check the configuration of your own computer and note the processor speed and RAM.

2 Find out as much detail as you can about your graphics card.

3 Locate the latest driver for your printer and, if you would like to, install it on your computer.

4 if possible, install an item of hardware.

5 Find out on your computer where you could plug in a USB memory stick/flash drive.

6 Change screen resolution temporarily and check the difference when the new settings are applied.

7 What capacity hard drive would you choose for a new computer?

   a 500MB

   b 700GB

   c 300KB

Printer settings

Most of the time printing one copy of the document or spreadsheet on screen will be all that is required. But what if you want something more? In that case, you will need to change settings. These changes could include what is actually printed, what print medium you want to use and how to correct printing errors.

Sometimes you will only want to change settings for a single print job, but you can also set these changes as the default so that all future printouts will be the same. To do this, you must go via the **Start** menu.

(The example used here is a Canon IP1800 printer, and your own printer options may appear differently.)

change printer settings

1 Open the printer dialog box from within a program, for example by clicking on the **Office** button and selecting **Print – Print**.

2 Click on the **Preferences** button.

   Or

3 Go to **Start – Printers and Faxes,** right click on the printer you want to use and select **Properties – Printing Preferences**.

4 Here you can make the following changes:

   a On the **Main** tab, select a different **Media Type** from the list, such as **Transfer** paper or **Envelopes**.

   b Set a **Print Quality – fast** will give a poorer quality printout which might be perfectly adequate for draft materials.

   c Click on **Greyscale** (black and white) to print without colour.

   d Click on **Colour Intensity – Manual** and then the **Set** button to change contrast and brightness.

Fig. 21.12 Print preferences

9

e On the **Page Setup** tab, select a different size of paper and whether to add borders (margins). If you select a paper size that is smaller than the Page Size, the page image will be reduced. If you select a paper size that is larger than the Page Size, the page image will be enlarged.

f You may have the option to print Poster size which will allow different parts of an image to be printed across several pages.

g To print on both sides of the page you can click on **Duplex Printing** if your printer offers this. If not, choose to print only odd pages from the normal Print dialog box, then when they have printed turn over and reinsert the batch of paper and print all even pages.

h For long documents printed on office machines, you can choose where staples will be added and whether to collate pages or reverse print order and print from the last to first page.

i You can also change page orientation at this stage if you did not do so when creating the document.

Fig. 21.13 Printer page setup

j On the **Effects** tab, click on **Monochrome** to print in a single colour – for example, Sepia.

k On the **Maintenance** tab, you can run a series of checks to make sure you are getting optimum performance and to correct errors, such as blocked nozzles, cleaning or realigning the print head.

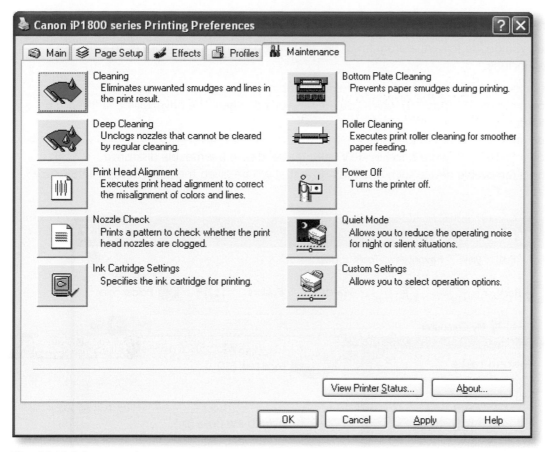

Fig. 21.14 Printer maintenance

Setting the default printer

If you are linked to several printers, you can choose which one to set as the default. Then, when you use a shortcut to print a document, the chosen printer will be selected automatically. It will also be the printer displayed in the **Name** box if you open the **Print** dialog box.

set the default printer

1 Go to **Start – Printers and Faxes**.
2 The printer showing a tick is the current default.
3 To change the default, right click on your preferred printer and click on the option to set this as the default.

Fig. 21.15 Default print

Saving files

When you decide to save a program or file, you have a wide choice of places to save it to. These include:

- the local C: drive or hard disk in a folder or subfolder or even on the desktop
- on a remote computer to which you are networked, where the drive will be labelled J: or S: etc.
- on an external hard or zip drive
- on a DVD/CD-R (write once) or -RW (rewriteable) disk in a writeable disk drive
- on a removable device such as a flash drive that will be given the next available letter such as E: or F:.

Fig. 21.16 Drives visible in my computer

For most of these, the process is identical.

1 Select a **Save** option to open the **Save As:** dialog box.
2 Navigate to the correct drive so that its name or letter appears in the **Save in:** box.
3 Name the file and select the appropriate file type.
4 Click on **Save**.

Fig. 21.17 Save As

write files to a CD

1 Insert a CD-R or CD-RW into the drive.

2 Select the files you want to save.

3 Click on **Copy these files** in the Tasks pane.

4 A balloon will appear on the taskbar stating that the files are ready.

5 Click on the balloon to open the disk window. Files to be saved will appear a paler colour at the top of the window, with any files already on the disk displayed below.

6 Click on **Write these files to CD** and follow the wizard to name the disk and copy the files across.

Click to start writing

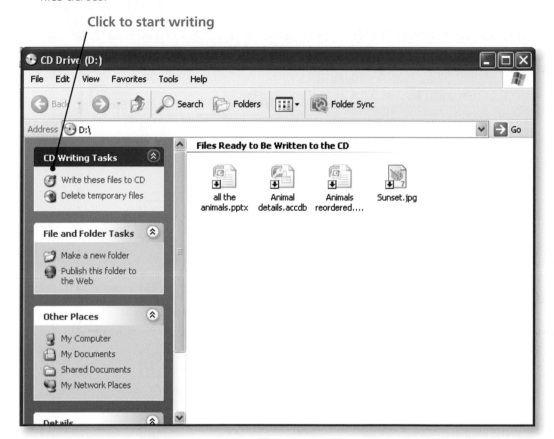

Fig. 21.18 Write to disk

Desktop shortcuts

One major reason for customising your computer is to speed up the time you spend working on it. A simple way to do this is to create shortcuts to programs or files you use regularly. This saves time searching for them and can be used on a temporary or permanent basis. (Although you can pin programs to the Windows XP or later Start menu for easy access and recently used programs will be added to the Start list automatically, you cannot do this with files.)

Once the shortcut is created, just double click on the icon to open the target file or program.

create a shortcut

1 Locate the program or file from the desktop – perhaps within **My Documents** or via **My Computer – Local Disk C:\ - Program Files**.

2 Make sure you can see part of the desktop on the screen.

3 Select the target file or program, then use your right mouse to drag its name onto the desktop.

4 When you let go, select the option to create a shortcut. (If you want to create a shortcut of a program on the Programs list, when you drag its name onto the desktop you will only have a copy option.)

5 An icon for the file or program will appear showing an arrow and named **Shortcut to…**

6 If necessary, rename it.

7 You can safely delete any desktop shortcut using the **Delete** key as it will only delete the link, not the actual file or program.

Fig. 21.19 Shortcut

Check your understanding 2

1 Set your printer to print a copy of a file in greyscale at the lowest quality onto transfer paper, but do not carry out the printing.

2 Run a test to check that the print head alignment is correct.

3 Save any file onto an external or remote drive.

4 Create a desktop shortcut to a program you use regularly.

5 Check which printer has been set as the default.

Networks

Computers that are linked together form a network and computer networks can link two computers together or thousands of computers across the country or even the world.

A small network such as one linking computers within a building or on a single site is known as a **LAN** (Local Area Network) and the most widely installed LAN technology is Ethernet. If the network connects computers across a city, they may be called **MANs** or Medium Area Networks.

Networks of computers communicating with one another across long distances are known as **WANs** (Wide Area Networks) and the largest WAN is the Internet. As there would be far too much cabling to install it all from scratch, the Internet makes use of existing communication systems such as standard telephone lines, TV cables and satellite dishes.

As well as the Internet, another type of WAN you may come across is an intranet. This is a private network normally run within an organisation that only staff and authorised customers or suppliers can access, which is used for company-related information and communication.

Whatever the distance between computers, there are two main types of network: **client-server** and **peer to peer**. In client-server networking, one system (the server) provides services for another (the client). The two systems usually reside on different computers and you may come across mail servers, web servers or database servers. In the peer to peer networks all the computers play the same role and no computer acts as a centralised server.

Protocols

These are the rules that govern communication on a network. The most commonly used are **TCP/IP** (Transmission Control Protocol and Internet Protocol). IP is concerned with dividing information into packets of data and getting it to the correct address. This is made possible because every computer and router on the Internet has a unique 32-bit address written in four groups of numbers such as 196.29.0.05. TCP is concerned with setting up reliable connections to deliver the data accurately and in the proper sequence across the network, as well as deaing with errors that might arise.

Other Internet protocols include **FTP** (File Transfer Protocol) governing the transfer of files across the network and **HTTP** (HyperText Transfer Protocol) controlling how web pages are written and displayed.

Network connectors

LANs work by sending signals to a central device, the networking switch, which directs traffic across the network. Signals are sent from a network interface card (**Nic**) inside each computer which identifies that particular machine with a unique address. Nics are connected to the switch by Ethernet cable or they use radio waves as their means of communication.

Communication across multiple networks can only work through the use of a device known as a **router**. This is a specialised piece of equipment that filters traffic across networks and ensures that data is forwarded or 'routed' to the correct destination. For example, routers enable email messages to reach a named recipient securely, without being seen by everyone else on the Internet.

Transmission speeds and signal degradation differ with the different connections and so, as well as cost, this will affect which one is chosen. Theoretical speeds may not turn out the same in practice as the speed often depends on range, traffic and other factors. For example, although you may be able to send large files using a wireless connection, it will be much slower than when using a wired connection because of congestion and interference.

A range of connections can be used and these have the following generally accepted transmission speeds:

- **firewire 400** (maximum 400 Mbps) and **800** (maximum 800 Mbps); firewire is also known by IEEE numbers: IEEE 1394, IEEE 1394a, IEEE 1394b, IEEE 1394c
- **USB** (12 Mbps) and **USB2** (480 Mbps)
- **cable modem** (1–6Mbps download)
- **wi-fi** (12–30 Mbps)
- **ADSL broadband** (1–2 Mbps download).

Wave formats

Different types of signal can be sent across networks.

- **Analogue signals** are in the form of an electronic wave where images and sounds are represented by continuously changing frequencies and voltage levels. This is the form of signal sent down normal telephone lines, which is why computer data sent this way first has to be transformed from digital to analogue using a modem.

- **Digital signals** divide images and sounds into the ones and zeros of computer language. The computer data provides a more consistent signal that is highly resistant to interference. Computers receive this data and decode it back into images and sounds.

- **Carrier signals** can be modulated to carry analogue or digital signals. They are transmitted at a steady frequency on which information can be imposed by increasing signal strength or varying other elements. Over optical fibres, a carrier can also be a laser-generated light beam on which information is imposed.

Types of computer

As you learned in Level 1, computers come in all shapes and sizes and their selection will depend on how you want to work or the job you want to carry out. The most common include the following.

Supercomputers

Supercomputers are among the fastest and most powerful machines used mainly for 'number crunching' or performing thousands of calculations. Their speed is measured in **tera** (10^{12}) or **peta** (10^{15}) **FLOPS** (floating point operations per second).

Mainframes

Mainframes are large computers designed by manufacturers such as IBM that are found in a wide range of organisations including banks and insurance companies. They are used particularly for bulk data-processing and financial transactions. You interact with a mainframe when you use an ATM to take out money. Their speed is usually measured in **MIPS** (millions of instructions per second).

Minicomputers

Minicomputers are mid-sized multi-user computers fitting between mainframes and today's PCs bought by small and medium-sized businesses at a much lower cost than a mainframe to run their business applications.

Microcomputers

Microcomputers are small stand-alone computers such as a PC that depend on microprocessors as the central processing unit. The term encompasses games consoles, mobile phones, pocket calculators, laptops and other familiar devices. The speed of these computers is measured in MHz and GHz, millions of cycles per second.

Workstations

Workstations are microcomputers used for a particular application such as scientific or technical work and so are normally optimised for displaying specialised or complex data. The term workstation is also applied to any computer, including a PC, that is attached to a network.

Laptops

Laptop computers (also called **Notebooks**) are portable computers the size of a small briefcase that contain an LCD screen in the lid and a touch pad instead of a mouse.

Pocket computers

Pocket computers are small battery-powered hand-held computers programmable in BASIC which have now been replaced by other hand-held devices known as **PDAs** (personal digital assistants) or palmtops which are stand-alone but can also link to desktop standard software. They usually have touch screens or make use of a stylus or even a miniature keyboard.

Mobile phones

Mobile phones can be used not only as conventional telephones but to access the Internet, take pictures, listen to music and send text messages (SMS). Popular manufacturers include Nokia, Sony Eriksson and Motorola.

Smart phones

Smart phones combine mobile phone with PDA technology.

Using applications

The development of a Graphical User Interface (**GUI**) has meant that almost anyone can learn how to work with software such as word processing or presentation packages very easily. This is because you choose from menus, drop-down lists and toolbars displaying understandable pictures (icons) rather than having to remember and type in code.

There are different ways to purchase and use application software. The three most common types available are:

- **stand-alone programs** such as an image editing program that has unique menus and tools – for example, Paint Shop Pro or Photoshop. These are valuable when you want to carry out highly specialised tasks.

- **integrated software** where a single package incorporates several different functions at a basic level such as a word processor, spreadsheet and database package – for example, Microsoft Works. These are particularly useful if you want to save money but still carry out a range of tasks. Often a basic integrated package will be included with a new computer. To get all the bells and whistles, you must buy the software suite.

- **software suites** that include a number of different programs offering advanced features that still work in a similar way and are compatible with one another – for example, Microsoft Office 2007. These are usually purchased when the advanced features are required for all the functions.

Cheap or free programs

Many cut-down versions of applications software are now available over the Internet for downloading or via magazine CDs. Some may be full but time-limited programs whereas others may include only the basic features but will give you a chance to see what the program can do. There are different types of program, including:

- **shareware** or 'try before you buy' – after the time is up, you will need to send in money to access the full features or keep the program running

- **freeware** or free software – genuinely free programs that you can use for as long as you like and may even be able to edit

- **abandonware** or **unsupportedware** – software that is no longer distributed or supported.

With any software written and distributed by others, there are certain legal restraints. Different terms you may come across include:

- **copyright** – resides with the creator unless it clearly states that the material can be copied or changed

- **copyleft** – describes the process of removing copyright restrictions on work so that authors can allow some copying or distributing of rights to users

- **licences** – needed to use software. The type of licence will limit it to being used by a single individual or across an organisation. Nowadays, it is often impossible to run software on different machines as you have to enter a registration key before it can work, and this can limit it to one installation

- **opensource** – relates to the access granted to the software's code so that developers can make changes.

Software developers often bring out new versions of existing programs, such as the Office 2007 suite. There are a number of reasons they do this:

- beating a competitor to market
- a competitor has released new software with additional features

- the previous version had been patched so often that an upgrade is the best solution
- the market demands new features.

There are no hard and fast rules about upgrading software – it is up to you to decide whether or not to buy the latest version of software you already use. Early versions of new software often contain bugs, but the manufacturer will fix these with patches sent as software updates.

Check your understanding 3

1 Is Microsoft Office Word 2007 part of an integrated software package or a software suite?

2 Is the intranet of a company with offices in several different cities likely to be a LAN or WAN?

3 Which protocol is involved when sending files from one computer to another?

4 Name three different types of computer.

Organising your files

You will already know from working at Level 1 that areas of the hard disk, remote drives or removable storage media labelled and set aside for storing related files are known as folders, and you will know how to create these folders. In summary:

1 Right click on the parent drive or folder on the desktop.
2 Select **New Folder**.
3 Name the folder that appears.
4 Repeat the process to create subfolders inside any folder.

Once a hierarchy of folders is set up, you can add files in two different ways:

- by moving or copying them into the folders from the desktop
- from within an application during a save.

move or copy files on the desktop

1 Open the folder in which the files are stored.
2 From the **File** and **Folder Tasks** pane, select the **Move** or **Copy** option.

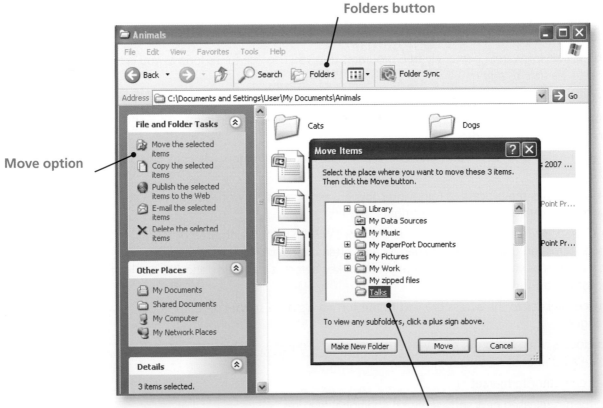

Fig. 21.20 Move files

3   Select the target folder from the list that will appear.

4   Click on **Move** or **Copy** and the files will be added to the folder.

    Or

5   If the files and destination folder are both on screen, use your mouse to drag the files into the folders.

    Or

6   If necessary, right click on the files and select **Cut** to move or **Copy** to copy them. Work through the folder pathway until you have opened the target folder and then select **Paste**. The files will appear inside.

    Or

7   Right click on the **Start** button and select **Explore**, or click on the **Folders** button to display the folders list in the left-hand pane.

8   Click on the folder in which the files are stored to display them in the right-hand pane.

9   Use the scroll bars to locate the target folder and make sure it is visible on screen.

10 Select the files you want to move (hold down **Ctrl** to copy) and drag them across to the target folder. When it goes blue, let go of the mouse and the files will drop inside.

11 Use the right mouse button so that, when you let go, you can select the **Move** or **Copy** option from the menu.

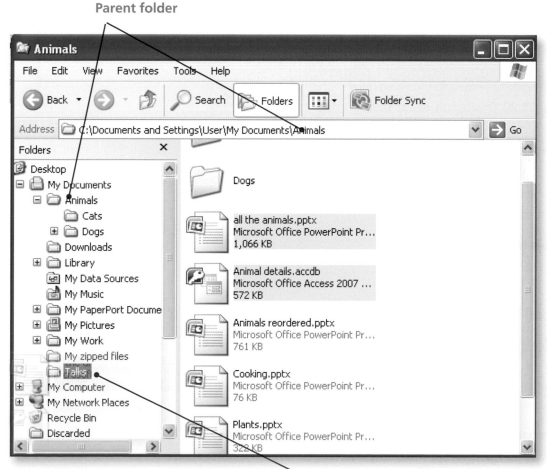

Fig. 21.21 Move folders using folders list   Target folder receiving the files

save files into folders within an application

1   Start to carry out a save in the normal way.

2   Work through the folder pathway until the target folder is displayed in the **Save in:** box

    Or

3   With the parent folder showing in the **Save in:** box, click on the **New folder** button to create a subfolder directly.

4 Double click on this folder to place it in the **Save in:** box and then continue to save the file.

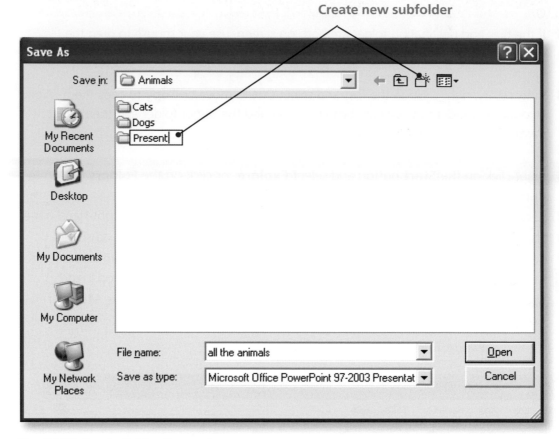

Fig. 21.22 Create folder for saving

rename a folder or file
1 Right click.
2 Select **Rename**.
3 Enter the new name.

delete a folder or file
1 Select it.
2 Press the **Delete** key.
3 Remember to move any files out of a folder first if you want to keep them.

Recovering deleted files

When you delete files or folders, they are not instantly removed but are moved to the **Recycle Bin**. On a network, this will be emptied regularly but you usually have time to restore files or folders if you have deleted them by mistake.

restore files
1 Open the **Recycle Bin** on the desktop.
2 Select the files or folders.
3 Click on the **Restore** option.
4 If you click on the **Empty** option, the files will be deleted permanently.

Restore   Permanent deletion

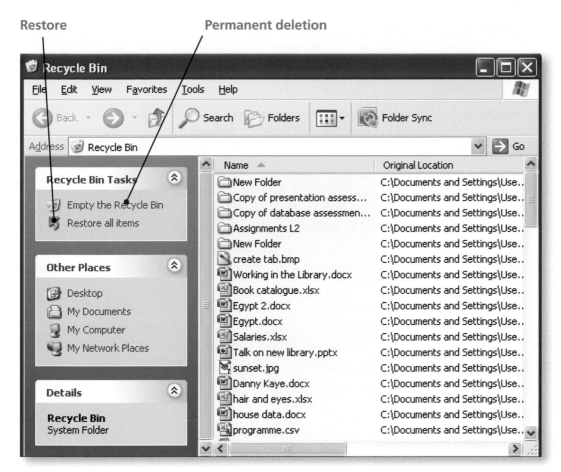

Fig. 21.23 Recycle bin

Check your understanding 4

1 Create a new folder named Location. This can be on your own computer (e.g. in My Documents), or on a remote drive.

2 Copy the file *Manchester* provided on the CD-ROM to this new folder.

3 Move the files *Liverpool* and *Paris* provided on the CD-ROM to this new folder.

4 Now create a subfolder inside Location named *Europe*.

5 Move the file *Paris* from your Location folder into this subfolder.

6 Take a screen print showing the contents of the Location folder.

7 Delete the files *Manchester* and *Paris*.

8 Restore *Paris*.

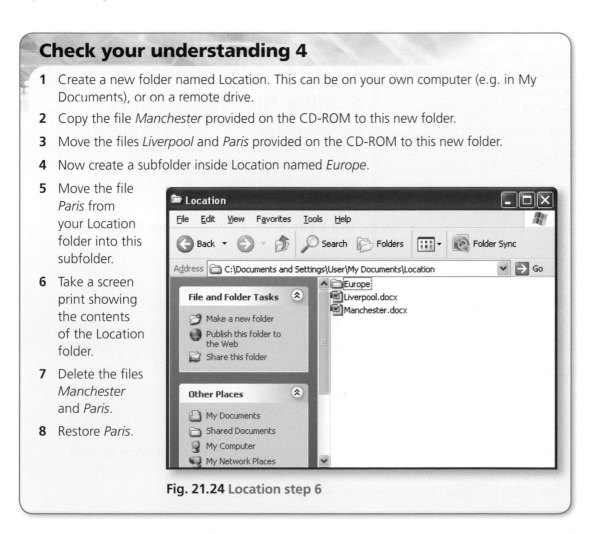

Fig. 21.24 Location step 6

File and folder attributes

As you will have noted, files and folders appear different when viewed on the desktop. Files display a small icon showing the file type or program in which they were created and folders appear as yellow boxes.

For any program in which you are working, there is a limited range of file types that you can create and save. These files are the ones that can be opened and read within the program. The file type is shown by its file extension, separated from the file name by a dot.

For example:

- graphics files may be **bitmap** (.bmp), **jpeg** (.jpg), **gif** (.gif) or **portable network graphics** (.png).
- word processed documents can be **plain text** (.txt), **rich text** (rtf), **97–2003** versions of Word document (.doc), **templates** (.dot) or **Microsoft Office Word 2007** documents (.docx).

One reason to change file type is so that it takes up less storage space. For example, plain text files are much smaller than Word documents as they contain no formatting, and archives (.zip) contain compressed files.

As well as selecting an appropriate file type you can also control whether it can be edited by others or even found at all. By changing its attributes, you can make it **Read-only** (otherwise known as **write-protected**), which stops anyone making any changes, or hide it from view.

- **Data files** are files stored by the computer and used by an application or system
- **System files** are used by the operating system to run your computer. In Microsoft, they tend to have the extension .sys.

Fig. 21.25 Properties

find out about a file

1 Right click on the file on the desktop.
2 Click on **Properties**.
3 You will see the type of file, which program it opens with, its size, when it was created and where it is located.
4 Click in the checkboxes to change any attributes.

Finding files and folders

Despite careful saving, it is common to lose track of a file or folder. You can usually find it using Windows' search facility, and the more details you have, such as the file name, size, likely location or date it was created, the quicker the search will be.

find a file

1 Go to **Start – Search**.
   Or
2 Click on the **Search** button in an open folder window.
3 When the Windows Search pane opens, click on the **Search Companion** – a small dog image.
4 Select a type of search or search all files and folders.
5 Enter part or all of the file or folder name, or any text in the file itself.
6 Click on any of the links to restrict the search further:
   a Select the most appropriate drive or folder location. Click on **Browse** to work through your folder pathway.
   b Select a range of dates between which the file was created.
   c Limit the size of the file to look for.
   d Search for hidden files or through system folders.

7 Click on the **Search** button.

8 Any files or folders relevant to your search will appear in the right-hand pane.

9 You can now open a file or folder that has been found, modify the search criteria or start again.

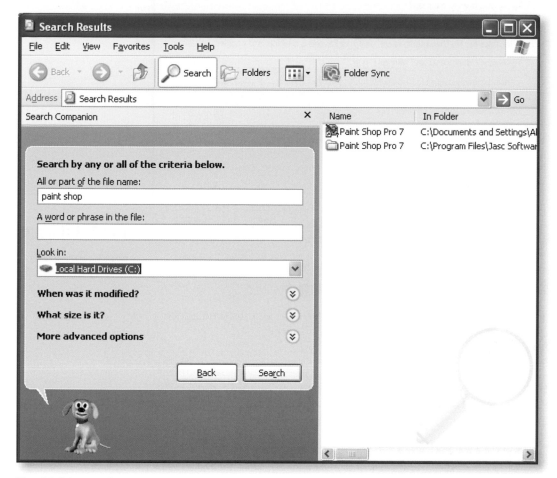

Fig. 21.26 Search

Making backup copies

To make sure you can always find an important file or folder, you should keep a copy somewhere separate from your own computer. This backup can then be used if the latest version is lost.

make backups

1 For individual files or folders, copy them onto removable storage media or a separate drive using the same procedure described earlier.

2 If you have Windows XP or later Professional, you can use the built-in backup utility to make copies of all your files and folders and even your computer settings, as well as schedule this to take place on a regular basis:

   a Go to **Start – All Programs – Accessories**.

   b Click on **System Tools**.

   c Click on **Backup**.

   d Work through the wizard by selecting what to back up and where to store the copies. Remember not to back up too many files if you won't need them again, and choose a location such as a networked or flash drive where there is enough space.

   e The material will all be saved in a single file.

   f Click on **Advanced** to schedule a regular backup, for example once a week.

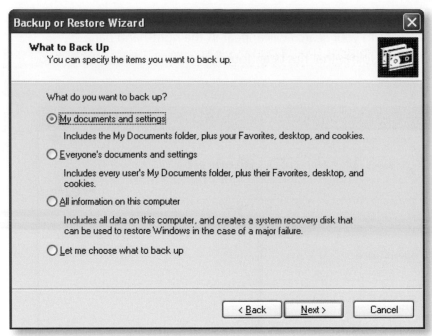

Fig. 21.27 Backup 1

Check your understanding 5

1 Use the **Search** facility to locate the file *Paris* you restored earlier.

2 Make the file Read-only.

3 Search for any jpeg image files on your computer stored in **My Pictures**.

4 Finally, search for any files that were created during the past week.

Working with windows

Each time you open a program into a window, you are presented with a standard range of toolbars. These house the menus from which you can select specific options, as well as a range of buttons that you click on to carry out a task.

To customise your windows, you can add or remove toolbars and also drag them into different positions.

change toolbars

1 Right click on any toolbar to see the list of available toolbars.

2 Names showing a tick are already on screen.

3 Click to take off a tick and remove an unwanted toolbar.
   Click to add a tick and add that toolbar.

4 To reorder toolbars, hold down the mouse button over the left-hand edge of a toolbar and, when the pointer becomes a four-way arrow, drag it below or above another toolbar.

5 For some programs, such as those that are part of the Microsoft Office 2007 suite, you will find that toolbars have been replaced by a ribbon with various tabs offering grouped commands. There is also a **Quick Access** toolbar for your most common commands. To add or remove commands from this toolbar:

   a Click on the down arrow at the end of the toolbar.

   b Click to add another button from the list, and reverse this process to remove unwanted commands.

   c Click on **More Commands** to open the full list of commands.

   d Select **Commands** in the left pane and click on **Add** to add them to the toolbar or **Remove** to take them off the toolbar.

Quick access toolbar

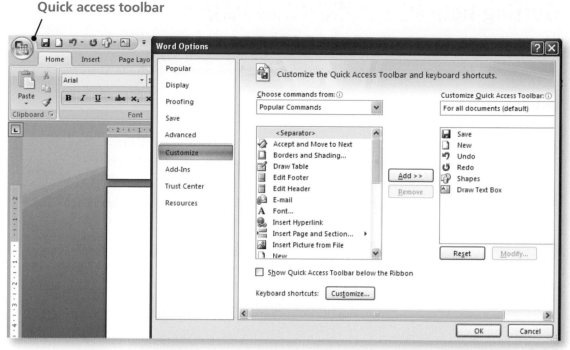

Fig. 21.28 Quick access toolbar

Carrying out tasks

There are various ways you can achieve the same goal when using a computer. You can:

- select an option from a menu dialog box
- click on a toolbar button
- use your keyboard.

Depending on circumstances, it can be easier or quicker to choose one method over another.

For example, to make selected text bold you can do any of the following:

- Open the **Font** dialog box and select the **bold** option.
- Click on the **B** toolbar button on the **Home** tab.
- Hold **Ctrl** and press the letter **B** on your keyboard.

IT problems

Unfortunately, even the best-maintained system can encounter problems. While it is impossible to list every problem you might encounter, some are more common than others:

- printer not working
- Internet unavailable
- program freezes / stops working
- save function disabled
- computer running slowly
- computer not running consistently
- error messages appear on screen
- computer does something incorrect or unexpected

Because the causes of the above problems can vary, it is not possible to list all of the possible solutions. However, there are some basic checks to keep in mind:

- Have all devices been plugged in? To the wall? To each other?
- Have all devices been turned on?
- Are your virus definitions up to date?

Most software packages have an integrated help facility which you can use to solve a range of basic problems. If you cannot find your specific query, try searching for "troubleshooting". It is very likely there will be a simple fix for common problems.

Getting help

Whatever task you want to carry out, you can get help in some form directly from your computer. There may be a link to the Internet where you can read guidance or watch demonstrations, or you can work through Help pages accessed within a particular program or from the **Start** menu.

get help

1 Go to **Start – Help and Support**.

2 When the Centre window opens, you can click on a topic and work down through the levels until you reach the required help or type keywords into the **search** box and click on the green arrow button.

3 Further topics will be listed in the left pane and clicking on any will display details in the right pane.

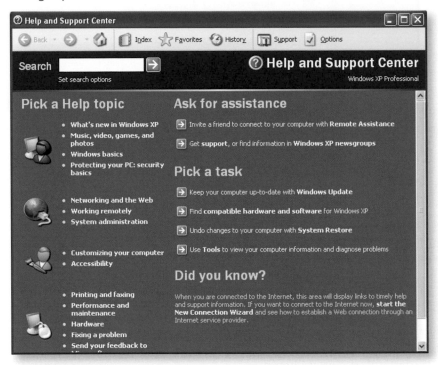

Fig. 21.29 Help 1

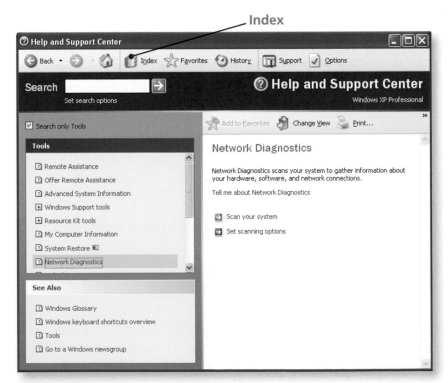

Fig. 21.30 Help 2

4 In many cases, you can click on a link to actually test your system and sort out a particular problem.
Or

5 Click on the **Index** button to display all the entries in **Help**. You can now search alphabetically or by using keywords.

6 Within a program, click on the question mark in the right-hand corner of a window
Or

7 Press the function key **F1**.

8 This opens the program's specific help screens that work in a similar way.

Error messages

At some time, everyone who uses a computer will find something goes wrong or they need to take action to avoid a problem. The computer cannot speak to you and so it communicates in the form of error messages. You should always read these very carefully and take the suggested action.

Messages will usually be one of three different types. They may:

- warn you before you make a mistake
- inform you about a mistake that has been made
- provide important information about the system.

For example:

- When saving, you may be warned that a file of the same name already exists, to prevent you overwriting it.
- You may be told that you cannot access a web page because you have typed the URL incorrectly.
- You may be told that you cannot shutdown properly as a driver needs updating.

Most error messages have a number and this can help you identify the problem if you search the **Microsoft Windows Knowledge Base** containing a database of all error messages.

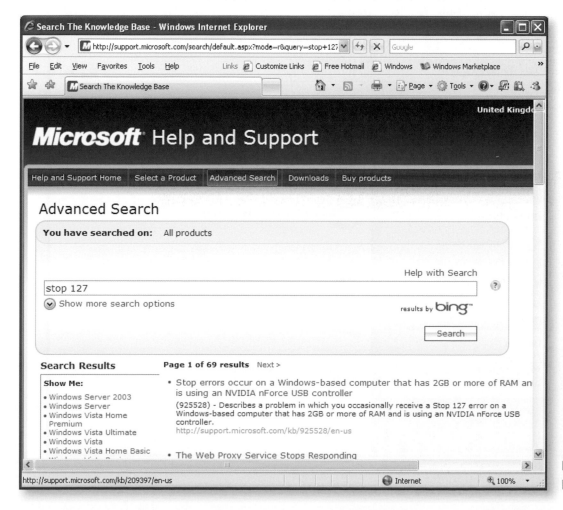

Fig. 21.31
Knowledge base

Accessing shared data

If you work in an organisation offering networked resources, you will be given rights by your IT administrator to access these. The administrator will determine:

- who can access shared data
- where users access this data
- what barriers are in place to prevent abuse of the resources.

You will probably have a 'home' directory where only you can view the files, as well as permission to access shared programs or files. These shared resources are likely to be on one or a group of servers and abuse will be limited through the enforcement of a strict User ID and password security system.

You will need to agree to abide by the code of conduct before being given access and if this is deliberately broken you may lose your job or be denied access to university or college facilities.

Sharing your own documents

When using Windows XP or later, you can allow other users access to your own files in various ways as you can make use of the simple file sharing facility that should be enabled automatically.

share files

1 Copy the files to one of the **Shared Documents** folders already set up by dragging to the link in **Other Places**.

2 These folders are counterparts to **My Documents, My Pictures** and so on which only you can access after logging in, and so this will enable anyone using your computer to work with them as well.

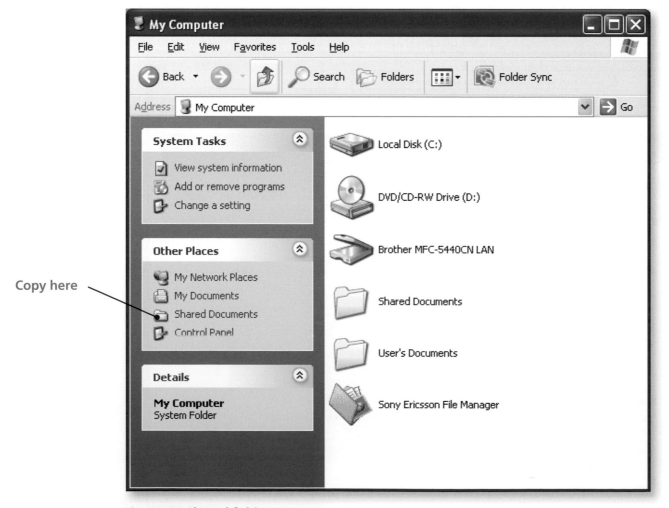

Fig. 21.32 Shared folder

3   On a network, locate the drive or folder you want to share.

4   Right click and select **Sharing and Security**.

5   On the **Network** tab, click on **Share this folder on the network**.

6   To grant others permission to modify these files, click in the **Allow network users to change my files** checkbox.

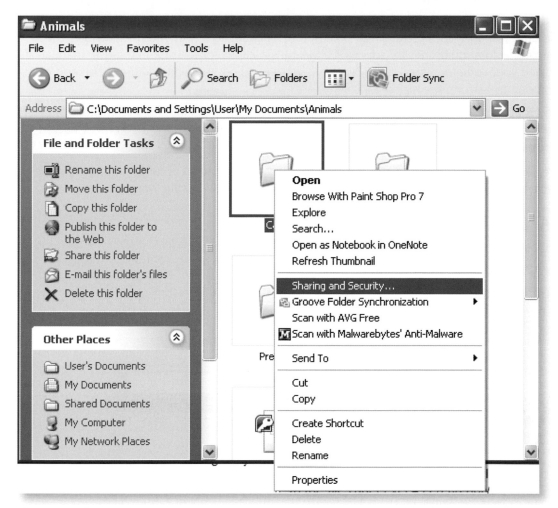

Fig. 21.33 Sharing

Check your understanding 6

1   Use **Help** to find out how to change your screensaver.

2   Make sure you know the rules governing the access of shared data within your organisation.

3   Note down any error messages you see over the next week and, if unsure, try to find out exactly what they mean.

4   Add two extra commands to the **Quick Access** toolbar in any Office 2007 program.

Using systems tools

As you save, edit and delete files and programs, the data on the hard disk will become scattered so that accessing files is slow. To correct this type of problem, you should regularly carry out a process known as **defragmentation**. This will move files so that they are next to one another and can be located more quickly.

You should also use **utilities** to check the hard disk for errors such as directory errors or bad sectors, and scan the computer for viruses.

Following these regular housekeeping tips will help optimise system performance.

Routine maintenance of hardware and software is a good habit to get into. Make sure to regularly defragment your hard drives, delete unwanted files and programs, and update software and drivers. It is recommended that you perform system housekeeping on a monthly basis, but if you regularly add new files to your computer – such as mp3s or photographs – a weekly defragmentation will help maintain performance.

Hardware maintenance is more difficult, as components are often covered by a warranty that is violated if you open the case or tower unit. Before maintaining any of your IT hardware, check the manufacturer guidelines for advice. Are there instructions for the task in the user manual? If not, and your computer is still under warranty, you should get the manufacturer or an approved technician to perform the maintenance.

Good routine maintenance habits are also key to user safety. Regularly check to make sure that your cables and plugs are in good condition, especially if there are pets in the home. Chewed wires are a hazard to you and your animals.

Non-routine maintenance is usually in response to a system problem or failure. These should be directed to a trained technician, especially if your computer is still under warranty.

defrag your hard disk

1   Open **My Computer**.
2   Right click on **Local Disk (C:)**.
3   Select **Properties** and click on the **Defragment Now** button on the **Tools** tab.
4   The hard disk will be analysed and, if necessary, defragmentation will take place.

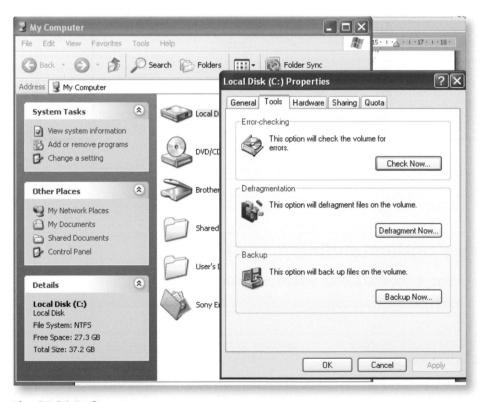

Fig. 21.34 Defragment

check the hard disk

1   Select **Local Disk (C:) – Properties – Tools** and click on **Check Now**.
2   Click on the **Start** button.
3   Click on **OK** when the **Complete** window appears.

Computer viruses are everywhere, but you can minimise risk of exposure by following some basic guidelines:

- Do not open attachments or click on links from people you don't know.

- Think when surfing – have you really been specially selected to win a Bugatti Veyron, or is there a reason that banner really wants you to click it?

- Not only is illegal file-sharing illegal, but P2P networks are filled with files claiming to be the latest movie that are really cover for a virus or trogan. Don't illegally download files.

- Check site security certificates if you are unsure.

- Regularly update your antivirus software and run full virus scans.

- Beware of data mining or spyware applications, and scan your hard drive for them regularly.

Fig. 21.35 Check disk

scan for viruses

1 Open your anti-virus software control centre.

2 Click to update the database via the Internet.

3 Click to start a scan.

4 Use the built-in scheduler to set a regular scan.

Problem	Troubleshoot by...
Program not responding	Wait for a minute to check the program is not just running very slowly. If it has frozen, press **ctrl + alt + del** and select Task Manager. On the Applications tab, select the frozen program and click End Task.
Paper jam	Modern printers often have displays with instructions relating to that specific device. If that is the case, follow the instructions on screen. If not, you must locate the jam and remove the piece of stuck paper. Once the jam has been cleared, printing will continue as normal.
Storage full	Clear unwanted files or programmes from your computer. You can move files to an external hard drive or other data storage devices, or simply delete anything you no longer want. To remove unwanted programmes, select **Control Panel** from **Settings** on the **Start menu**. Double-click on **Add or Remove Programs**. Choose the program you would like to remove, and then click **Remove**. Repeat until you have removed all of the programs you no longer want.
Error dialogue	Follow any instructions that appear within the error dialogue. If no instructions appear, make a note of the error code and search the **Help Centre** for instructions relating to that error code.
Virus threat	If your antivirus software detects that an incoming email or attachment has a virus, delete it immediately. Do not attempt to open it. If a virus threat is found during a system scan, you will need to delete or quarantine the affected file. Instructions will vary, so check your manual or help files for advice.
Memory low	Conduct some routine PC housekeeping. Consider archiving unwanted files by moving them to disks or an external hard drive, and run a disk cleanup and defragmentation once space has been created.
Connection loss	Check that all of your cables are plugged in, and check the power on your router. If you have power and your cables are all plugged in, try resetting your router. Instructions vary between devices, but a common fix is to unplug the router from the power for 30 seconds before plugging it in and turning it on.

Verifying desktop configuration

You will soon have problems if you have not adjusted the sound correctly, the computer's date and time are incorrect, the mouse click is too fast or you are using the wrong regional settings for the dictionary, spelling or other features. All these settings can be changed from the **Control Panel**.

Fig. 21.36 **Configure**

change settings

1   Go to **Start – Control Panel**.

2   Double click on any option to display its properties.

3   Use sliders, checkboxes, pointers or other facilities to amend the settings.

Check your understanding 7

1   Carry out a disk cleanup and defrag.

2   Change the regional and language options to US English and note the differences this makes to the samples before changing back to UK English (or your local language if you are out of the UK).

3   Check that the time is set correctly.

4   Change the screensaver running on your computer.

Health and safety issues and good practice

When working with computers, it is important to follow common-sense procedures. These include:

- looking after the equipment
- taking care of yourself
- reporting and correcting faults and hazards.

The computer

Computers can be damaged by a sudden increase in power so it is worth fitting **power surge protectors**. These prevent your computer from being damaged by a sudden voltage increase caused by a dramatic power surge, such as lightning.

As with all electrical equipment, you should take care to keep liquids and debris away from the machine and to keep it clean. Monitors, storage media, keyboards and so on should all be cleaned regularly with proprietary cleaners and cloths. Such materials should be stored carefully and only used according to the manufacturers' instructions.

Personal safety

The three areas where problems can arise if you spend long periods working on a computer are:

- physical strain on your back, arms and hands
- problems with your eyes
- general fatigue.

It is easy to feel safe on the internet – while surfing you are in your home, school, or office, somewhere you feel safe. But the internet is not a safe place, and cyber criminals abound.

In the real world, you would not give out your personal or banking data – or that of your friends and family – to any stranger who asked for it. You would not meet a stranger without putting adequate safety precautions in place. Apply those real world rules to the online environment.

To avoid Repetitive Strain Injury (**RSI**), brought about by sitting too long in one position or under strain, the most important advice is to take regular breaks and make sure you change position frequently.

There is a variety of ergonomic equipment available that is designed specifically for computer users. This includes chairs that support your back, wrist rests for your hands, specialised types of mouse and footstools to keep you sitting comfortably while maintaining a good posture. You can also use screens to cut down on glare from the monitor, or magnifiers if you find it hard to read screen contents.

Changing the settings on the computer can also help. For example you can adjust contrast and colours to make it more restful as you work, and increase text size on Web pages.

It is important to set up your work area so that everything you need is within easy reach rather than requiring you to stretch awkwardly, and stop working if you have a headache or feel particularly tired.

Identifying and reporting faults or hazards

If you notice any fault in the equipment you are using or you see something that you feel is potentially hazardous such as an overloaded power supply or bare wire, it is important that you report it. This may be by contacting a network administrator by phone or email, or by filling in a report card.

There are different levels of hazard that you may come across, and you will need to carry out a risk analysis to decide on the appropriate action. These include hazards that are:

- **passive or dormant** – they have the potential to cause harm but no one has yet been affected
- **active** – they are certain to cause harm as nothing can stop them
- **armed** – they are ready to cause harm but you can avert trouble by taking action.

There are some basic guidelines to follow for safe and secure IT use:

- Protect your personal data to prevent fraud and identity theft.

- Be careful when using images – some may cause offence or break laws including copyright.

- Respect confidentiality – yours and others'.

- Make sure you use appropriate language. This can apply to the task at hand – an email to a friend is different to a job application – or to the clarity of your messages

- Use the Bcc field where possible to prevent sharing email addresses.

> ## Remember
>
> Regular computer housekeeping – archiving files, making back-up copies, and defragmenting when finished – will keep your computer running at peak performance.

Check your understanding 8

1 List three items of computer equipment you should always clean regularly.

2 What term is used to describe equipment that helps you work healthily and safely?

   a   Economic

   b   Ecogenic

   c   Ergonomic

3 Is a coat left lying on the floor near the computer a passive, active or armed hazard?

4 What equipment protects the computer from sudden high voltages?

5 Which two pieces of advice would you choose to give a new computer user to make sure they worked safely and healthily?

Assignment

This practice assignment is made up of four tasks

- Task A – Systems specification
- Task B – Folder structures
- Task C – Keyboard shortcuts and default settings
- Task D – Good working practices

You will need the following files:

- Answer Sheet.docx
- L2-7267-021.zip (containing: Errors document.docx)

Scenario

You work for a local charity shop that sells donated clothes and other goods raising funds to provide Christmas and birthday presents for local orphaned children. The shop has been given a computer system and you have been asked to produce a document detailing the system specification, so the charity has a basis for identifying possible future software and hardware upgrades, and to set up the system.

You may use your own computer system to represent the donated system for these practice assignment tasks.

Please read the text carefully and complete the tasks in the order given.

Task A – Systems specification

1   Working on screen, use the specifications of your computer to complete the **Answer Sheet** document provided on the CD-ROM accompanying this book.

2   The charity has an old Hewlett Packard HP PSC 920 printer that was recently donated (but not yet delivered) and you are to install the driver for it.

Ensure the printer defaults with A4 paper in portrait at the most economical setting.

Take screenshots of these settings.

3   You have been asked to suggest an additional USB printer. It is intended that this printer will be capable of producing photographs and printing from pen drives.

Use a search engine to select a suitable printer. Enter the manufacturer's URL on your document and take a screenshot of the web page.

Download the latest driver for your selected printer to My Documents. Take a screen print while the transfer is in progress.

Task B – Folder structures

1   Using wildcard and date search techniques, search your computer for all documents modified in the last two days. How many files did you find?

Take screenshots of this.

2   Create a new folder named **Documents backup** in **My Documents.** Copy the documents you found into the **Documents backup** folder.

Take screenshots of this.

3   Unzip **L2-7267-021.zip** from the CD-ROM accompanying this book.

Copy and paste the folder structure from this both into your **Documents backup** folder AND to a pen drive.

Take a screenshot of this.

4   Find the folder named **Move this** and move it to the **Documents backup** folder.

Take a screenshot of this.

5   Find the folder named **Rename me** and change the name to **Renamed**.

Take a screenshot of this.

6   Find the folder named **Delete me** and delete it.

Take a screenshot of this.

7   Find the folder you made named **Documents backup,** then look at the properties to find out the size on disk of the document files in this folder.

Take a screenshot of this.

8   Open the recycle bin to recover the deleted **Delete me** folder and undelete it.

Take a screenshot of this.

9   Find the folder named **Hide me** and set the attributes to 'Read Only' and 'Hidden'. Apply these changes to the folder, subfolders and files.

Take screenshots of these settings.

10 Use backup software to create a full backup of the **L2-7267-021** folder into your **Documents backup** folder. Call the backup **Assessment01**. Save the log of the backup operation to your **Documents backup** folder as **Assessment01.txt**.

Take screenshots of this.

Task C – Keyboard shortcuts and default settings

1   Use the Windows help system to find out how to specify shortcut keys for specific programs.

Create shortcuts for the programs on your computer you use the most.

Take screenshots of these settings.

2   Open the document called **Errors document** from the **Errors document** folder. Attempt to print this document. There will be an error message; respond to the message to correct the error.

Take a screenshot of how you corrected the error.

3   Save **Errors document** into the shared documents folder on your computer. If you have problems locating this folder, use the Windows help system to either locate the folder or find out why it's not present.

Take a screenshot to show it saved into the shared folder OR write a short explanation of why it's not there.

4   Adjust the mouse double-click speed property to suit yourself.

Capture screenshots of the visual prompts in the dialog box to show a double click has been recognised.

5   Check your computer has been set to UK for the region, currency, date and keyboard.

Take screenshots of these settings.

6   Check your computer has the correct date and time.

Take screenshots of these settings.

7   Mute the volume on your computer.

Take screenshots of these settings.

8   View your desktop display settings to find the screen resolution and colour depth.

Take screenshots of these settings.

9   Open your Internet browser and take a screenshot of it.

Customize the browser by adding two extra buttons of your choice.

Take a screenshot of the modified browser.

Task D – Good working practices

1　Start the Disk Defragmenter, but do not actually run it (to save time).

　　Take a screenshot of this.

2　Create a shortcut to your **Documents backup** folder on the desktop.

　　Take screenshots of this being created.

3　Show the properties of a local disk then start the error-checking tool.

　　Take a screenshot of this running (or any error message it produces).

4　Start a virus scan of the computer.

　　Take a screenshot of this then cancel the scan.

5　You have noticed that the wires are loose in the mains plug of your computer.

　　Please indicate on your **Answer Sheet** document what you would do about this situation.

6　On your **Answer Sheet** document, write an explanation of safe working practices when using computers regarding:

　　a　Eye sight safety
　　b　Ergonomics for using a computer workstation
　　c　Lifting & handling equipment

7　Write a short explanation on your **Answer Sheet** document of how files can be recovered if they have been accidentally deleted.

8　Write into your **Answer Sheet** document which of the following will allow the fastest data transfer speed:

　　a　USB 1.1
　　b　USB 2.0
　　c　IEEE Firewire

9　Save your **Answer Sheet** document inside the **Documents backup** folder named **Answer Sheet RJM** (where RJM = your initials).

　　Take a screenshot of this.

10　Use your word processor's zoom options to see a whole page at a time.

　　Adjust the page breaks and sizes of screen for your screenshots so each page of your document has three screenshots with their task references.

　　Use the print preview as a final check before printing.

Word processing software

This unit concentrates on producing business documents using a range of facilities that include merging documents with mailing lists, adding data imported from elsewhere, introducing consistency to the layout of the pages and improving the appearance of any tables.

At the end of this unit you will be able to:

➔ plan, prepare and produce new documents

➔ use mail merge facilities

➔ edit existing documents

➔ check documents

➔ save and print.

Business documents

As you saw at Level 1, a wide range of documents can be created using word processing software such as Word 2007. Each type of document is distinctive in terms of its layout, paper size and formatting and it is important to create documents according to the particular requirements or 'house style' of the organisation for which you work.

Types of business document

Common types of business document include the following.

Business letters

Business letters normally begin with the sender's address, perhaps in the form of a letterhead, followed by the recipient's address and the date. They start *Dear Sir or Madam* (or *Dear Mrs X*) and this is usually followed by the subject of the letter. They finish with *Yours faithfully* (or *Yours sincerely* if the recipient's name is known). Such letters are normally printed on A4 paper that is in portrait orientation, i.e. short sides top and bottom.

Memos

Memos are internal communications and normally have a top section showing four entries:

To – the name of the recipient

From – the name of the sender

Subject – the topic of the memo

Date

The main body of the memo then begins below this header.

Forms

Forms can be created offering boxes that people complete by ticking or writing in answers.

Reports

Reports are often very long documents that start with a main title and details of the writer. To make them easy to use, and in case pages get out of order, they should have a contents list, numbered pages and possibly brief details of the subject on every page.

Newletters

Newsletters are commonly divided into columns, like newspapers, with pictures used to break up the text. They can go over several pages and stories are usually broken up so that later sections are continued on inner pages.

Invoices

Invoices are sent to customers or clients and show details of goods or services provided together with what is owed.

Fax cover sheets

Fax cover sheets have space for the names of the sender and recipient, fax numbers, the date and the subject of the fax. It is a good idea to include the number of pages being sent, in case these become separated when the fax is received.

Advertising flyers

Advertising flyers are often smaller – perhaps printed on A5 paper. They can be in portrait or landscape orientation (with long sides top and bottom), and may include borders, differently sized and enhanced text and graphics to increase their attractiveness.

Brochures

Brochures are multipage documents, usually coloured, containing full details of the item or items for sale – for example, a brochure for the sale of a country house.

Leaflets

Leaflets are often produced to provide information about specific topics such as medical advice or surgery details.

Itineraries

Itineraries display details of journeys and may include timetables, locations, hotels and other travel arrangements.

Questionnaires and surveys

Questionnaires and surveys will contain lists of questions that may have answers to write in or numerical grades to select.

Paper size

Paper sizes vary considerably and it is usually your personal choice which you use. The default setting in word processing applications is A4, but this may not be appropriate for many documents. If you are using different sized paper, make sure you have placed the correct sized paper in the printer tray and checked and changed the settings for your printer before taking a hard copy.

Common paper sizes in the UK are:

A3 = 42cm x 29.7cm – used for drawings, posters and large tables
A4 = 21cm x 29.7cm – used in offices for letters and forms
A5 = 21cm x 14.8cm – notepad size
A6 = 10.5cm x 14.8 cm – postcard size

> ## Check your understanding 1
> 1 Try to find examples of different business documents.
> 2 Compare their style, layout, paper size, margins and orientation so that you are familiar with the differences.

Creating new documents

After launching Word 2007, you are always offered a blank document. Here you will have basic settings – the defaults – such as font type, font size, margins and orientation pre-set, but these can be changed to suit your work. However, you need to start designing the document from scratch.

For many business documents, it is quicker to customise a standard layout provided by one of the many templates within Word. Normal documents using these layouts can be created and saved in this way with the underlying template remaining unchanged.

create a document based on a template

1 Open **Word**.
2 Click on the **Office** button.
3 Click on **New**.
4 In the window that opens, click on **Installed Templates** in the index on the left.
5 This displays a range of document types. Click on any one to see a preview.
6 Select your preferred template and then click on **Create**.
7 When the template opens, customise it to suit your needs. In some cases, there will be boxes to click into and you can delete and retype any entries.
8 Save it as a normal Word document.

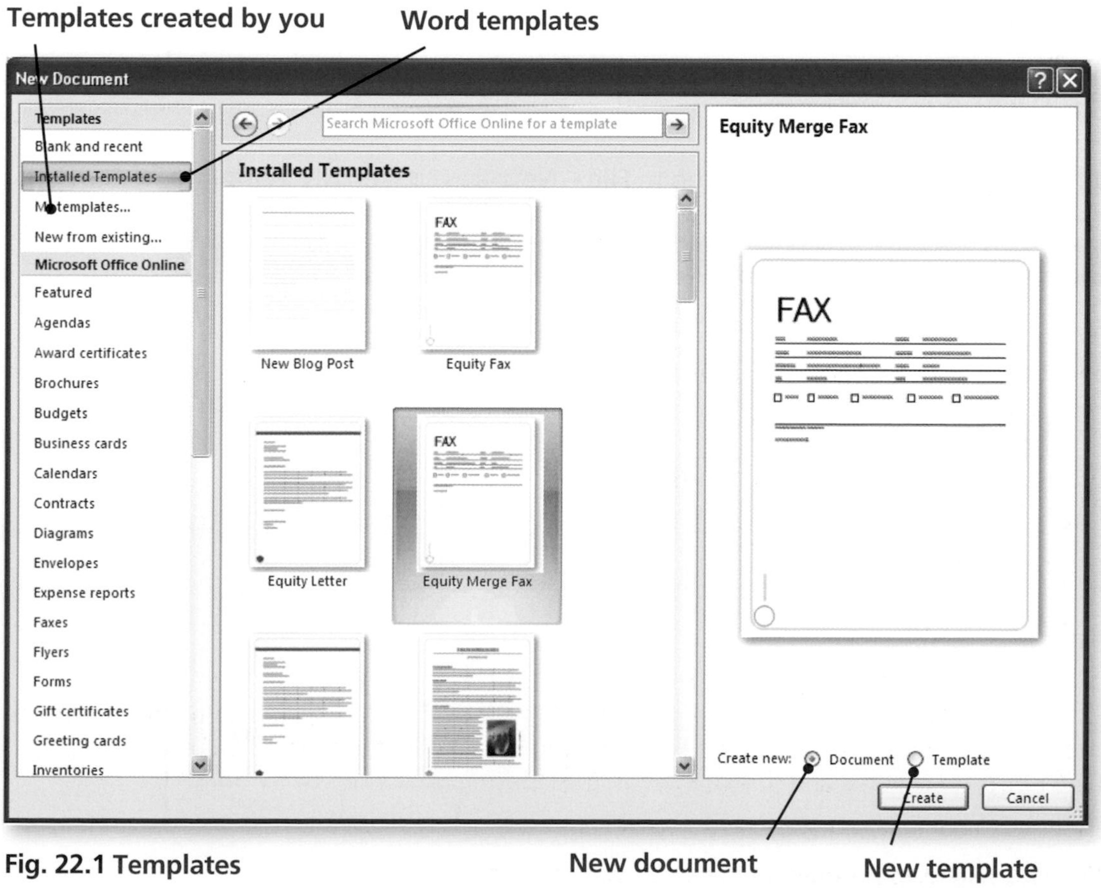

Fig. 22.1 Templates

You can also create your own templates that you can use whenever they are needed. They can include just a page layout or those text entries that will always be included.

create a template

1 To base your new template on an existing one, locate the template as above.

2 Click in the **Create new: Template** radio button.

3 Click on **Create**. A new document will open labelled Template1.

4 Format and change the layout and content as necessary.

5 Click on **Save** and you will open the Templates folder.

6 Rename the template and click on **Save** to store your new template with others in the **My Templates** section of the folder. You can also select a preferred folder location.

   Or

7 Start with a word processed document and design a template from scratch. Here, make sure you select a **Template document type (.dotx)** when saving.

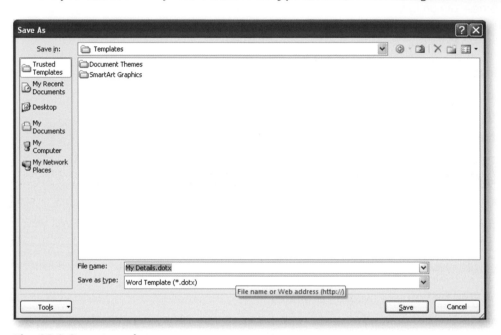

Fig. 22.2 Save template

8 To use it for future documents, go to **File – New** and click on **My Templates** in the index.

Fig. 22.3 Use own template

9 For templates stored elsewhere, open them as normal but make sure you change the file type to a Word document when saving any new document.

Opening files created previously

There are two basic methods for opening files:

- Locate and open a named file from the desktop.
- First open a program such as Word 2007 and then use the facilities within it to locate and open the file.

Locating and opening a named file from the desktop

With the file name visible, double click and it will open into an appropriate program with which it is compatible and has been associated. For example, a **workbook (.xlsx)** and **comma delimited file (.csv)** will normally open into **Excel** and a **bitmap** image file into **Paint**.

If you want to open a file using a specific program that is not set as the default, you can create a new association.

change which program opens a file

1 Right click on the file name on the desktop.
2 Select the **Open With** option.
3 Select a program from those listed if one of these is appropriate.
4 To open the file with a different program, click on **Choose Program**.

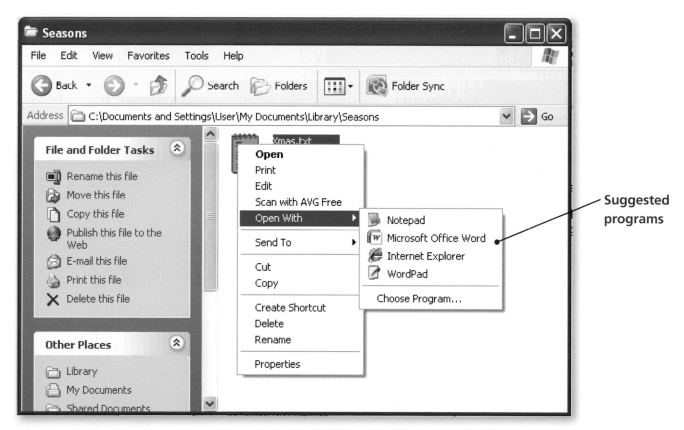

Fig. 22.4 Choose program

5 In the new window, select your preferred program from those available.
6 Click on **OK** to open the file.
7 If you want to use the selected program when opening all future files of the same type, click in the box labelled *Always use the selected program*.

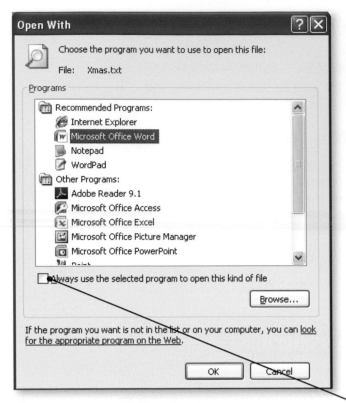

Fig. 22.5 Open with list of programs

Suggested programs

Locating and opening a named file from within a program

open a file within an application

1 Launch the program, for example Word 2007.

2 Click on the **Office** button.

3 Click on **Open**.

4 If the target file is listed in the Recent Documents pane, click on its name. Otherwise click on the **Open** button. (You can also hold down **Ctrl** and press **O**).

5 When the **Open** window appears, search for the folder containing the target file.

6 For files on a different drive, such as on a CD or networked server, click on **My Computer** and then open the correct folder to search inside.

7 Double click on the file or click on the file name and then click on the **Open** button.

Note that when using this method to locate different types of file, you may need to change the *Files of type* to *All files*, or select a specific file type, for example **.txt** (Plain Text file) or **.rtf** (Rich Text File). Otherwise you will only be able to search for Word 2007 documents.

Locate files on other drives

To open specific types of file

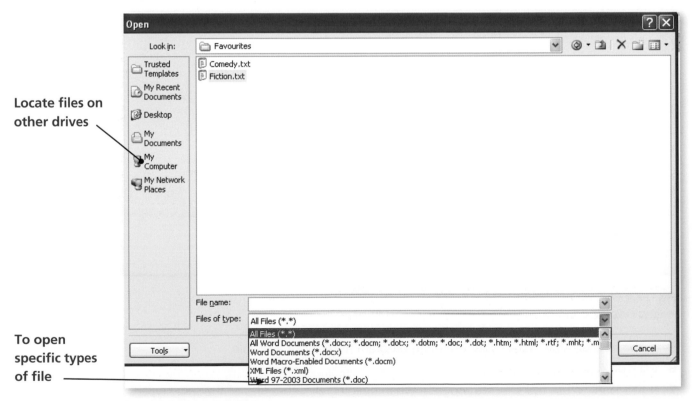

Fig. 22.6 Opening files

8 You may be offered a **Convert file** box, so select the correct file type and then click on **OK**.

When you use a particular software application there is a limit to the types of file you can open. If you try to open one of the wrong types – for example an image file using a word processing application – you will see an error message saying it cannot be read, or strange symbols and a box offering the chance to convert the file into one that is readable.

Some types of file are known as **generic files** as they can be read by a number of different programs. These include simple text files such as **.txt** and **.rtf** files, or data files such as **.csv** (comma delimited).

Fig. 22.7 Convert file

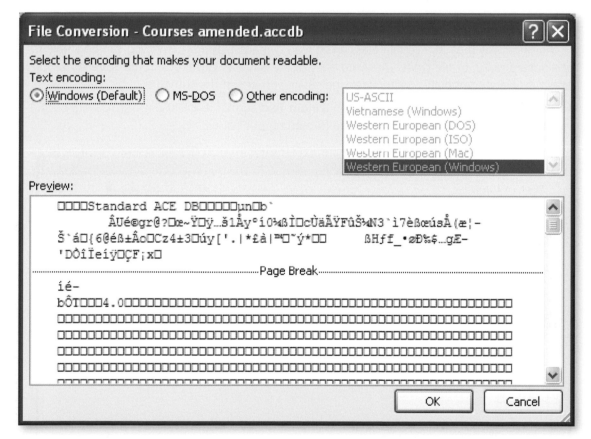

Fig. 22.8 Wrong type of file

Check your understanding 2

1 Open the image *Sunset* on the CD-ROM accompanying this book into the default image editing program on your computer and then close it.

2 Now open it using a different image editing program, for example **Paint, Paint Shop Pro** or **Windows Picture** and **Fax Viewer**.

3 Finally, open **Word 2007** and then try to use the normal open facilities to open the image file *Sunset* within this application.

Screen prints

To display new settings or show that folders have been created or accessed, you may need to provide evidence in the form of screen prints. Although covered at Level 1, here is a reminder of the steps to take.

take a screen print

1 Press the **Print Screen (PrtSc)** key on your keyboard for a picture of the whole screen.
Or
2 Hold **Alt** as you press the key to print only the active window.
3 Open a new document or image editing program and press **Ctrl + V** (or select **Edit – Paste**).
4 The screen image will appear.
5 Save and/or print the file in the normal way.

Note that when carrying out an assessment, it is a good idea to create and save a file with a name such as *Screen prints* and keep this minimised. Whenever you need to take a screen print, open the file and paste in the next image.

Saving

In the same way that only certain files can be read by particular programs, you can only save files in a limited range of file types. These are listed in the **Save as type:** box in the **Save As** window and may include templates, web pages and earlier versions of the software.

Note that if you save a Word document in a **Plain Text** format it will lose its formatting. Saving in a **Rich Text Format** will maintain some formatting but not all.

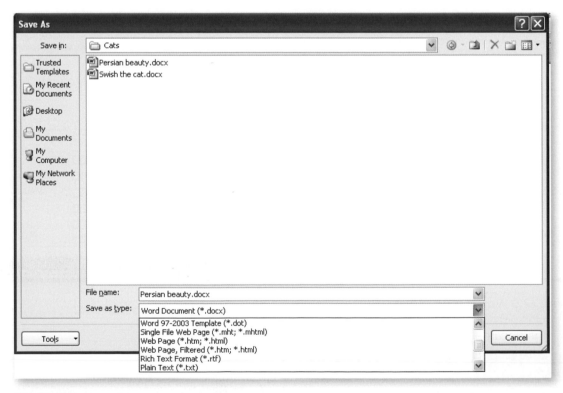

Fig. 22.9 Save as type

Printing

Before printing any documents, take the following precautions:

1 Proof read and spell check carefully on screen.
2 Check in print preview to make sure documents have no obvious spacing or other errors.
3 If necessary, change page orientation, margins or paper size at this point.
4 Click on **Shrink One Page** if a small amount of text has just gone over onto a further sheet.

Layout options

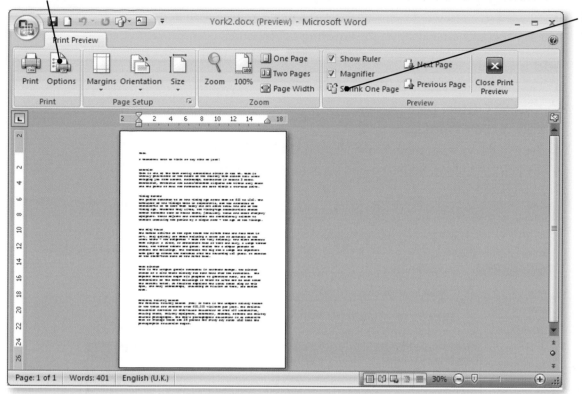

Fit text on fewer pages

Fig. 22.10 Print preview

5 Open the **Print** dialog box and change any settings if the number of copies, pages or appropriate printer has not been selected.

6 Check the printed output and, if necessary, correct the document on screen and print again.

use the spell checker to check a whole document

1 On the **Review** tab, click on the **Spelling and Grammar** button.

2 Make sure English (UK) is selected as the dictionary language.

3 When the first word is highlighted, click on an **Ignore** option to retain it in the document, or change it as follows:

  a Click on a suggested alternative in the **Suggestions** window

  Or

  b Click into the red highlighted word and correct it manually.

4 Click on a **Change** button to update the document and replace the word once (**Change**) or wherever it appears (**Change All**).

5 For a word you will use repeatedly, click on the **Add to Dictionary** button so that it is accepted in future.

6 When the entire document has been checked, click on **Close** to close the window.

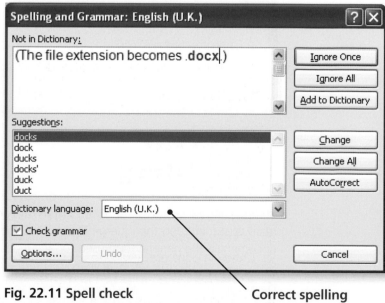

Fig. 22.11 Spell check

Correct spelling

Adding to the dictionary

If you want to add a word to the dictionary so that it is recognised and not flagged up in future, you can also add it manually via the **Proofing** option.

add a word

1 Go to **Office – Word Options – Proofing**.
2 Click on **Custom Dictionaries**.
3 Click on **Edit Word List**.
4 Type in the word in the box and click on **OK**.
5 To remove a word, select it in the list and click on **Delete**.

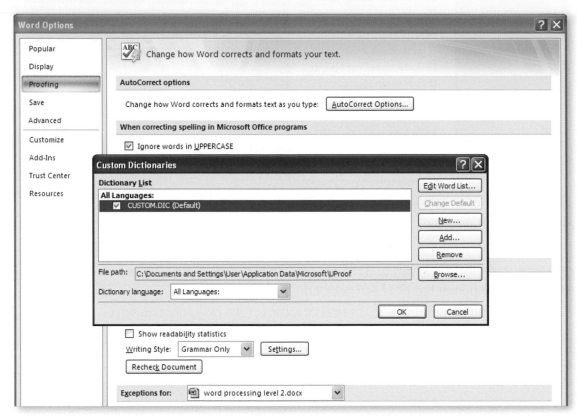

Fig. 22.12 Custom dictionary

Check your understanding 3

1 Open **Word 2007**.
2 Open the text file *York* on the CD-ROM accompanying this book.
3 Save it as *York2* in a Word 2007 document format. (The file extension becomes .docx.)
4 Check the document carefully and print one copy.

Mail merge

Instead of sending personalised letters or invoices individually to a large number of people on a mailing list, you can use Word's **mail merge** facilities. This allows you to create a single main document and then merge this with the details of everyone on the mailing list. The main document is the general term used for any type of document such as a letter, invoice or mail shot. You can use the same facilities to create labels or envelopes.

Data from the mailing list will be inserted in the form of **fields** – in other words the headings under which the data is stored, for example Surname, Town, Postcode etc.

The completed main document might start like this:

\<First Name\> \<Surname\>

\<Address Line 1\>

\<Town\>

\<Postcode\>

Dear **\<First Name\>**

When merged, each letter will include details within the specified fields drawn from a separate record in the database.

LETTER 1	LETTER 2
John White **3 The Willows** **Manchester** **M28 4LP** Dear **John**	**Mary Sheldon** **26 Ash Road** **Worthing** **BN13 4EP** Dear **Mary**

As well as names and addresses, the data source can include other information about the individuals if this will be needed in the final, merged documents – for example the library books they have borrowed, their salaries, dates they are starting a course, birthdays etc.

carry out a mail merge

1   Create or open the main document

2   Create or locate the data file containing all the names, addresses and other personal information. This can be in the form of a word processed table, database file or spreadsheet, as long as it is set out appropriately. This means that it must have column headings (**field names**) in the first row and one data entry for each field.

3   Merge the documents.

4   Print your merged documents directly, or save them in the form of a new, single merged document known as a **form letter**. This will have all the letters set out on a new page with each letter displaying one person's details. These merged letters can be stored separately from the source data file.

Creating the main document

If you want to use a document that is already created it is important to remove all personal details such as names and addresses as these will be drawn from the data file. Or you can start by simply opening Word and creating the document directly.

create a main document

1   Open Word and go to **Mailings – Start Mail Merge** on the ribbon.

2   Select **Step by Step Mail Merge Wizard** to open a help pane alongside the screen.

3   Step 1 – select **Letters** for any type of document other than labels or envelopes and then click on **Next**.

4   Step 2 – click on *Use the current document* if no letter has been created.

5   Click on *Start from existing document* to open a saved document on screen.

6   At this stage, you could type in any details that will be included in all the merged documents, or leave this until later.

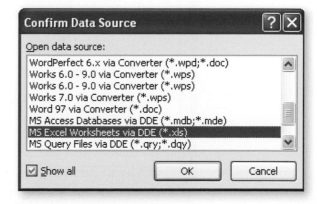

Fig. 22.13 Step 2 mail merge

Accessing the data source

Word 2007 offers the option to create a special Office database file during a mail merge. But it is common in many organisations to use data already stored on a large database elsewhere. You can therefore choose whether to create the data source file from scratch or link your main document to an existing one.

access an existing data source

1 Click on **Next**.

2 Step 3 – to link to an existing database, select this option and browse for the file. As you may need to look for a non-Word file such as an Excel spreadsheet or Access database, the system is set to search for 'all data sources' automatically.

3 When located, select the file in the main window and click on **Open**.

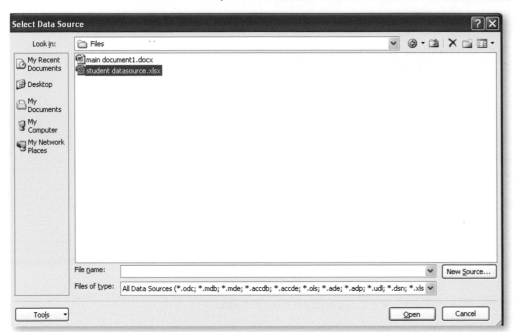

4 You may be asked to confirm certain details, such as the specific sheet of an Excel Workbook on which the data is held.

5 When you click on **OK** you will return to the main document and the name of your data source file will be displayed in the wizard pane.

Fig. 22.14 Browse for data source

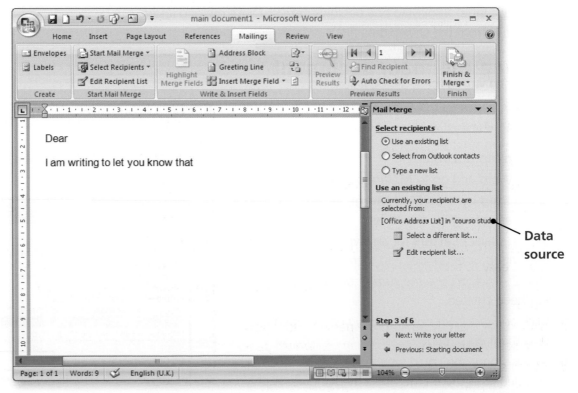

Fig. 22.15 Data source file in wizard

> ## Check your understanding 4
>
> **1** You are going to carry out a mail merge. Use the document *Lunchtime* as the main document.
>
> **2** Select the file *Lunchtimedata* as the data source.

create a new data source

1 At Step 3 click on **Type a new list – Create**.

2 A **New Address List** window will open displaying a range of headings.

3 Type the details of the first person on your mailing list into the appropriate boxes in the first row and then click on **New Entry** to start typing in record 2.

4 Continue entering records until all the details have been added.

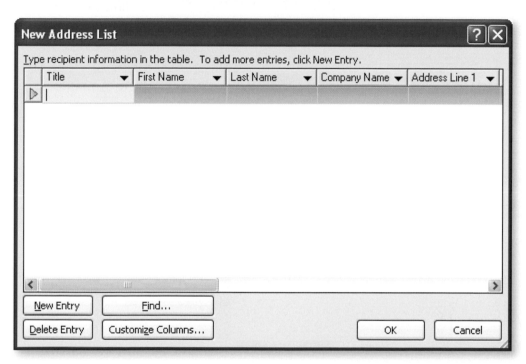

Fig. 22.16 New address list

5 To create your own fields/column headings, click on **Customize Columns**. You will be presented with a list of all the fields.

    a Click on **Delete** to remove unwanted fields from your data source file.

    b Select a field and click on **Add** to add a new one *below* it in the list.

    c Click on **Rename** to amend the field name.

    d Click on **OK** when the list is complete.

6 When all records have been entered, save the data source with an appropriate name and in a location that is easy to find. (Note that by default the file will be saved into the **My Data Sources** folder in My Documents.)

7 Back in the main document, the name of your new data source file will be displayed in the wizard pane.

8 At any stage, as long as you are not engaged in a mail merge, you can open the data source file to add or amend records.

Fig. 22.17 Customize headings

Merging data

Once you have linked your main document with a data source, you can start merging the two files. This involves inserting the correct field (sometimes referred to as a merge code) from the data source into the document at an appropriate place.

insert fields

1 Click on **Next** to start writing the letter.

2 Step 4 – you can now start or complete the document by inserting fields as appropriate.

3 Click in the place in the letter where you want to add details drawn from the data source file and then click on **More items**.

4 Select the field and click on **Insert** to add it to the document.

5 Click on **Close** to close the **Insert Merge Field** window and continue writing the letter.

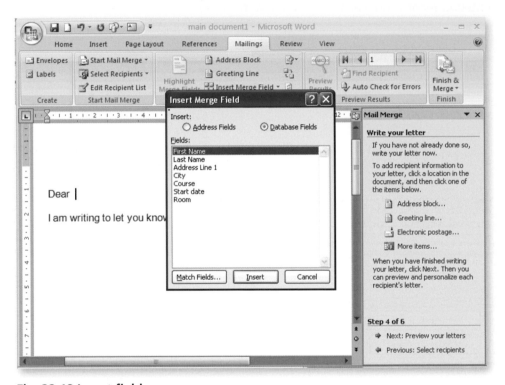

Fig. 22.18 Insert fields

6 Where relevant, use the shortcuts offered:

    a Click on **Address block** to add the various address details in your preferred format.

    b Click on **Greeting line** for a *Dear...* entry.

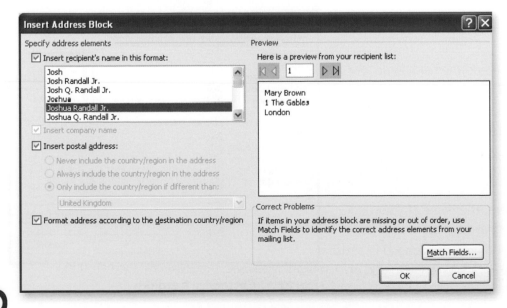

Fig. 22.19 Address block

7 When complete, your document will contain a mix of field names and normal text. The fields will be within << >> chevrons.

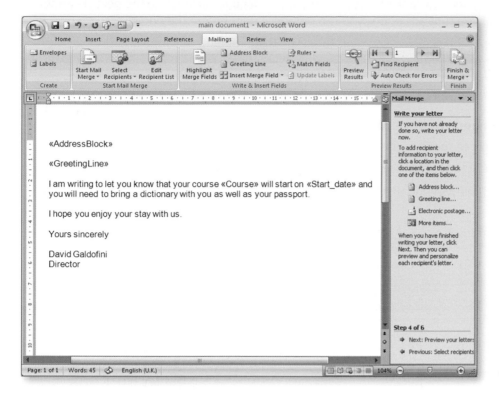

Fig. 22.20 Finished document with fields

Check your understanding 5

1 You are going to complete setting up the *Lunchtime* main document so, if you have not yet done so, first carry out Check your understanding 6 to link the document to the *Lunchtimedata* source file.

2 Insert all the fields in the appropriate places within the memo.

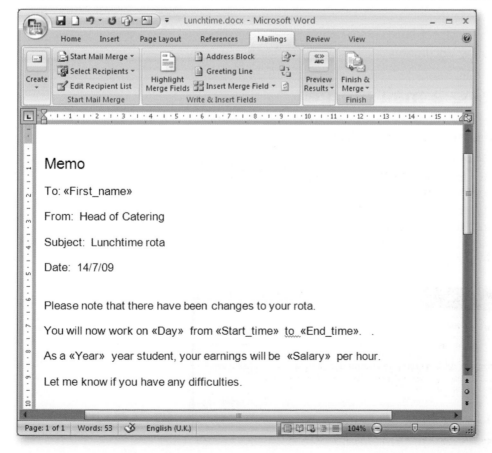

Fig. 22.21 Lunchtime merge1

Reviewing your documents

To see what the letters will look like when printed, you need to move on to the next step and preview them. If you spot any errors, you can put these right before the documents are printed.

At this stage, use normal formatting tools to select an appropriate font type and size, set the margins and page orientation and make sure text does not stretch just over onto a new page. One common mistake is to insert fields without the right gaps between them. This usually only becomes clear when you preview the letters with actual names and addresses inserted.

preview merged documents

1 Click on **Next** to move to Step 5.

2 Move through the records by clicking on the **Next** and **Previous** buttons.

3 Format the main document and correct any errors.

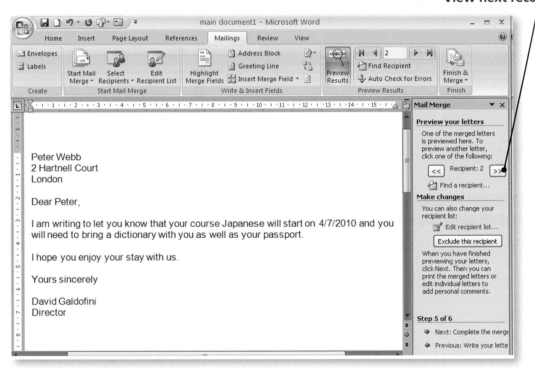

Fig. 22.22 Preview documents

Completing a mail merge

Once the letters have been completed, you have three choices:

- Print them directly.
- Save them as a single merged document temporarily titled *Letter1*, with each individual letter displayed on a separate page. As this duplicates the data, this alternative should normally be chosen only when you need to access or print the merged documentation separately from the data source.
- Save and close the main document and data source files without printing. When you open the main document later, it should be linked to the data source file. To continue with the mail merge, choose the **Yes** option.

Fig. 22.23 Opening main document after closing

print merged letters

1 Click on **Next**.

2 Step 6 – click on the **Print** option and decide whether to print all or selected letters or just the current document open on screen.

3 The **Print** box will open and you can change settings and print as normal.

4 Save the main document if you want to access it again in the future.

create a new document

1 At Step 6, select the *Edit individual letters* option.

2 Include all or some of the records and produce a new, merged document.

3 To make the merged document easy to find in future, save it with a name other than the default.

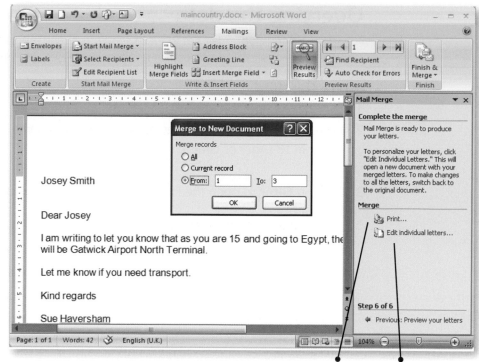

Fig. 22.24 Print merged letters   **Click to print**   **Create merged document**

Check your understanding 6

1 Following on from Check your understanding 7, preview the memos and correct any spacing or other errors.

2 Open the memo addressed to Nalia on screen.

3 Print a copy of this current document only.

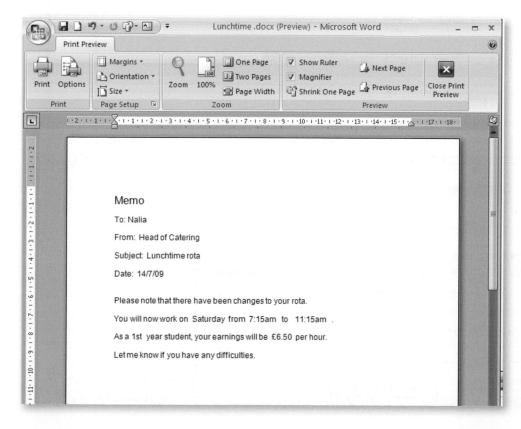

Fig. 22.25 Printed merged letter

Querying the data

For some mailings, you may not want to send out all the letters to everyone in the database. You can therefore select recipients before printing or merging your documents.

query the data

1 Click on the **Edit Recipient List** button.

2 Click on one of the drop-down arrows in the field name box to select a single entry, for example 18 (Age) or Greece (Destination).

   Or

3 Click on **Filter** to open the **Filter and Sort** window. You can now complete the following:

   a **Field** – select the appropriate category, for example *Age*.

   b **Comparison** – choose from the options such as *less than, equal to, more than* etc.

   c **Compare to** – enter the criterion on which to filter, for example *less than 18*.

4 Click on **OK** and continue with the mail merge. The records will only include those meeting your chosen criteria.

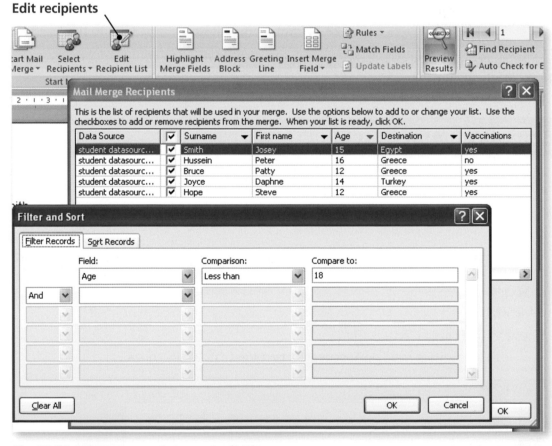

Fig. 22.26 Select merge data on 1 criterion

Using the mail merge ribbon

When you click on the **Mailings** tab, you will see a range of mail merge buttons along the top of the screen. You can use these to carry out a mail merge directly, without working through the wizard.

1 Click on the **Start** button to select the type of main document.

2 Click on **Select Recipients** to browse for the data source file or create a new one.

3 Filter out particular records by clicking on the **Edit Recipient List** button.

4 Insert fields using the **Insert Merge Field** button, or add a full address or greetings entry.

5 Click on **Preview Results** to see the merged letters, and click on the arrows to move backwards and forwards through the records.

6 Click on **Finish** to print or merge the documents.

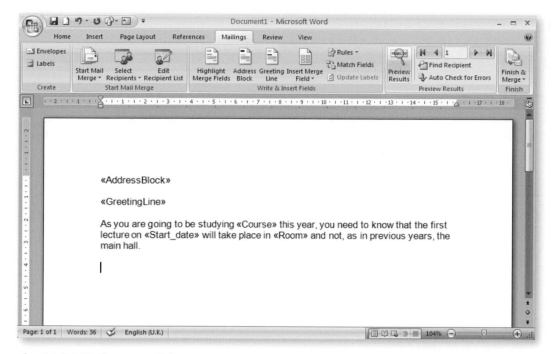

Fig. 22.27 Mail merge ribbon

Check your understanding 7

1 Start a mail merge using a new blank document.

2 You are going to create a new recipient list. Remove all unwanted fields, add the four listed below and then enter the following six records:

Florist	Date	Town	Number
Floribunda	September	London	250
Flowers for all	March	Manchester	180
West London Flowers	August	London	310
Floribunda	August	Exeter	156
Say it with	October	Norwich	320
Hazels	June	London	455

Table 22.1

3 Save the data source file as *Plantdata*.

4 Create the following main document and insert the fields where shown:

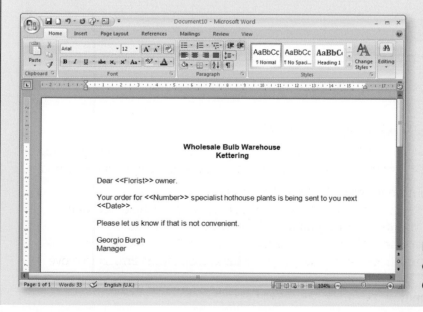

Fig. 22.28 Example of florist main document

5 Print only the merged documents for orders going to London (there should be three).

6 Save the main document as *Plant mailmerge*.

3 records selected →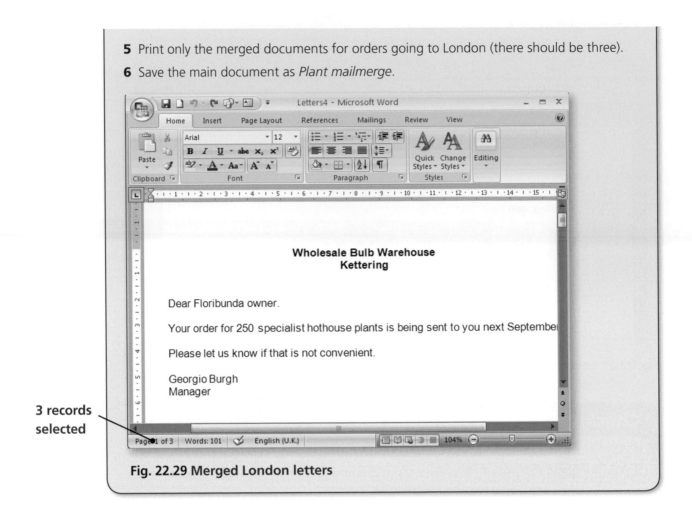

Fig. 22.29 Merged London letters

Creating labels or envelopes using mail merge

The steps needed to produce badges, mailing labels or envelopes are just the same as those used to create a mail merge document.

create labels or envelopes

1 Start a mail merge. It is quicker to use the wizard.

2 Step 1 – select **Labels or Envelopes**.

3 Step 2 – Click on the **Options** button. When the window opens, select the manufacturer and exact size of label or size of envelope. Click on **New Label** to enter your own measurements if your own label is not shown.

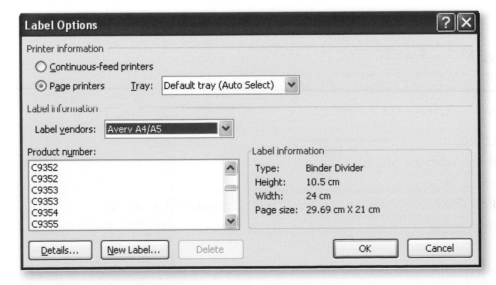

Fig. 22.30 Label mail merge

4   Step 3 – select recipients in the normal way.

5   Step 4 – for labels, the page should now show a range of labels and you can click on **Address block** to add these details to the first label or use the **Insert Merge Field** button to add your own choice of entry.

6   Click on the **Update** button to set out all the labels in the same way.

7   For envelopes, the page size will change. Click for a dotted box and insert the address and other merge fields as appropriate.

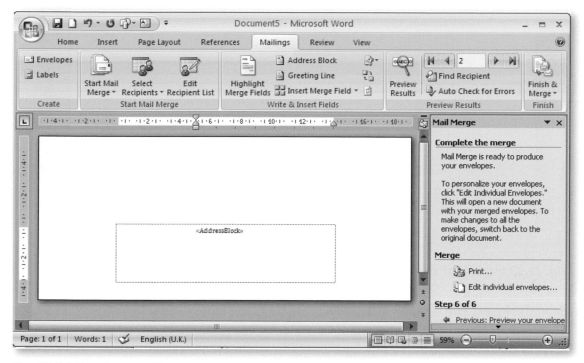

Fig. 22.31 Envelope mail merge

8   Preview the labels or envelopes and print or save as normal.

Integrating documents

It is often necessary to import data into business documents that has been created and saved already. Word enables you to bring in text, images, graphs and other data very easily.

The simplest method is to use **Copy** and **Paste**, but in some cases it is better to use the insert function. Insert options vary, depending on the source of the data you are introducing, and you must take care that the imported item is displayed correctly and does not, for example, extend into the margins or get split across pages.

Inserting text files

When you import text documents, the text will retain its original formatting and you can then leave this or apply the main document formatting as appropriate. If the imported file contains any objects such as images, these will also appear.

insert a text file

1   Click where you want the text to appear.

2   Click the **Insert** tab.

3   Click the drop-down arrow next to **Object** in the Text group on the ribbon.

4   Select **Text from File**.

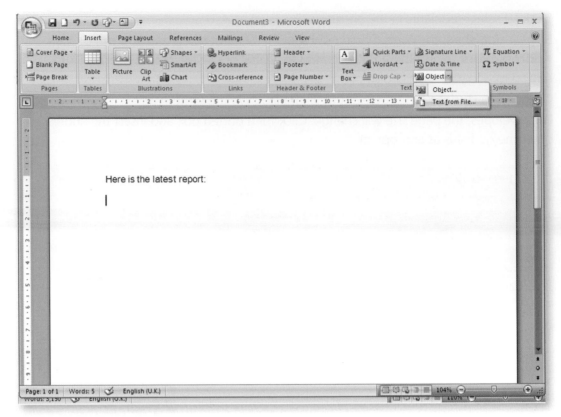

Fig. 22.32 Insert text file

5 In the **Insert File** window that opens, navigate to the file you want to insert. When looking for files other than Word documents, for example **rich (*.rtf)** or **plain text (*.txt)**, first change the *Files of type* to *All Files* or select the relevant type.

6 Select the file name in the window and click on **Insert**.

Type of file being inerted

Fig. 22.33 Insert file window

7 You will return to your main document and the new text will have been added.

8 If necessary, correct the text spacing. It may help to turn on the **Show/Hide** button that reveals all hidden formatting symbols.

Check your understanding 8

1 Open the file *Sportsday* on the CD-ROM accompanying this book.

2 Click on the line below the words *who won a prize* and insert the contents of the file *Results*.

3 Enter the name *Charlie Foxton* at the end of the document.

4 Save as *Integrated*.

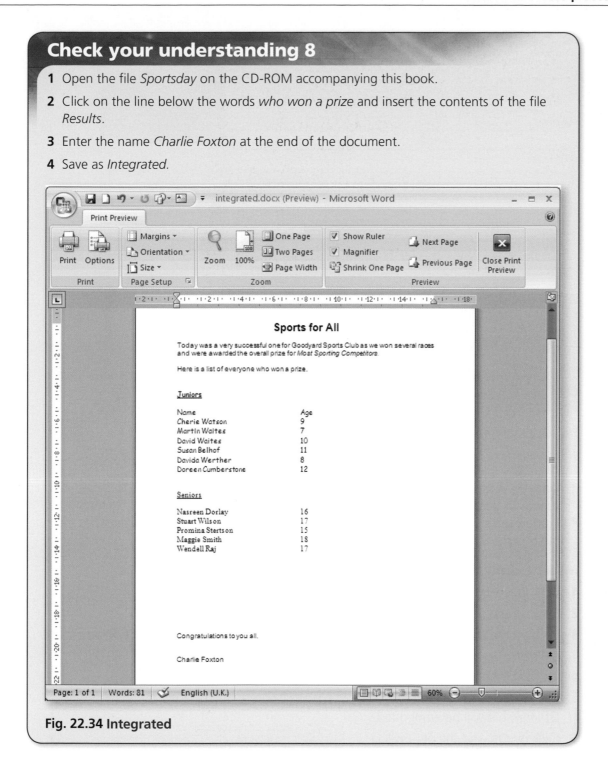

Fig. 22.34 Integrated

Importing data files

Information that you want to import into a Word document may be held in something like a spreadsheet or CSV file. When imported directly into a Word 2007 document it will normally appear as a table, but the editing options are different from normal Word tables. This is because Word uses **OLE (Object Linking and Embedding)** technology so that, when double clicked, you will be able to use formatting tools from the original application. You will also be able to format the object to a certain extent with Word tools.

Note that exactly the same process is involved if data is stored in a PowerPoint file or you want to import a chart.

Word 2007 converts most imported data into a table, but if it appears as text, you can use the facilities to convert it into a normal Word table. For example, a CSV file opened with Excel appears as a spreadsheet but in Word can display text entries separated by commas.

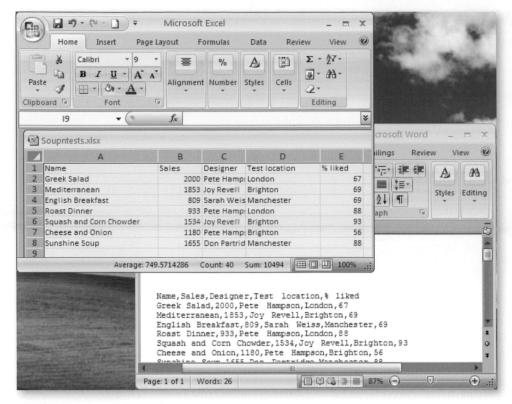

Fig. 22.35 CSV and Excel

There are two ways data files can be imported:

- **Linked** – if the original data is changed, you will be able to update the data imported into your main document as it is stored in the source file.
- **Embedded** – here, you can only change the data in the main document manually as the data has become part of the Word file.

insert a data file

1 Click where you want the data to appear and go to **Insert – Object**.

2 Click on the drop-down arrow and select **Object**.

3 Click on the tab labelled **Create from File** and then click on **Browse** to navigate to the data file.

4 Click on **Insert** and its name will appear in the **File name** box.

5 Click in the **Link to file** checkbox if you want to keep the imported data up-to-date. Otherwise it will be embedded.

6 Click on **OK** and the data will appear.

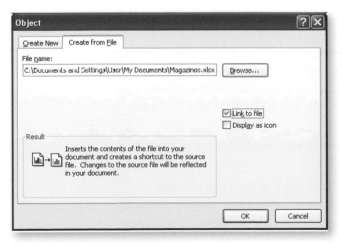

Fig. 22.36 Insert data file

Fig. 22.37 Imported data file

edit imported data

1 Double click on the table to edit it.

2 For an embedded (non-linked) table, you will display all the tools to allow you to format the numerical data, insert or delete cells and perform calculations.

3 Scroll bars will enable you to move round the sheet and you can click on a sheet tab to change to a different sheet if the data is elsewhere in the workbook.

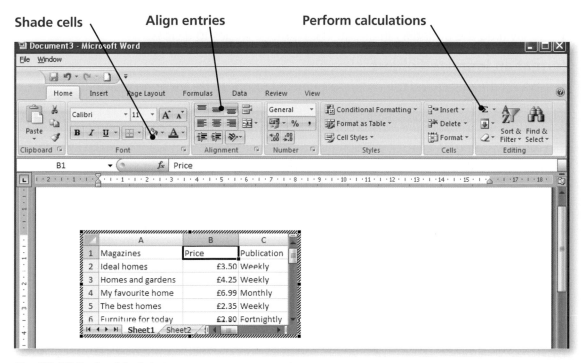

Shade cells   Align entries   Perform calculations

Fig. 22.38 Embedded data

4 For a linked chart, you will open the original spreadsheet in a program such as Excel and can use all the facilities for formatting or editing the data.

5 Return to Word by closing the application.

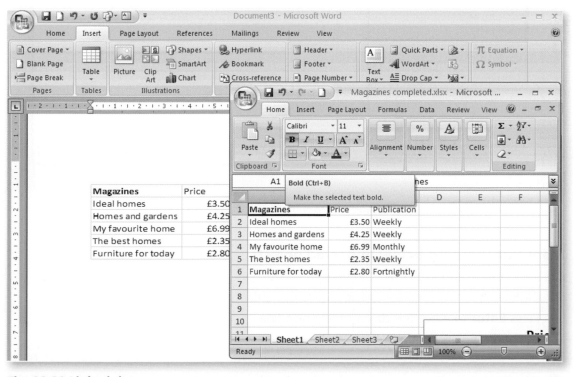

Fig. 22.39 Linked data

delete an imported object

1 Select the object with one click.
2 Press the **Delete** key.

resize an imported object

1 Click on the object once to select it.
2 Move the pointer over a corner (to maintain its proportions) until it shows a black double arrow and then hold down the mouse button.
3 Gently drag the boundary in or out. The pointer changes to a cross.

format an imported object

1 Click on the object and use normal alignment buttons to centre or right align it on the page.
2 Right click and select the **Format** option to open the dialog box or use tools on the **Home** and **Page Layout** tabs. Some of the changes you might want to make include setting an exact size, adding borders and setting a text wrap for text wrapping round it on the page.
3 Note that some options may not be available for imported objects.
4 If a text wrap is set, you will be able to drag the object to a different position on the page.

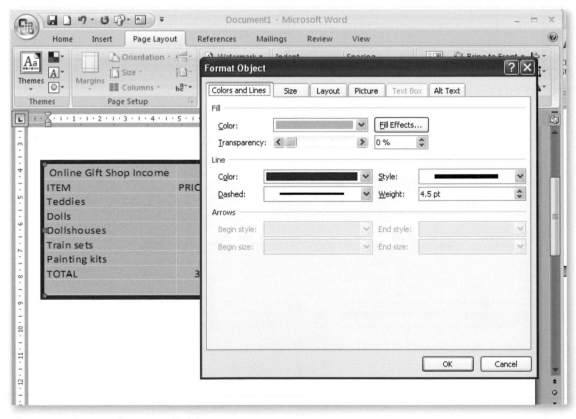

Fig. 22.40 Format imported object

continue working below imported objects

1 Click on the imported object to select it.
2 Click again and the cursor will appear to its right.
3 Press **Enter** to move the cursor onto a new line.
4 Continue typing as normal.
   Or
5 Double click on the page below the object to display the cursor.

Check your understanding 9

1 Start a new document and save as *Problems*.

2 Type the following:

   This year we saw a range of problems with the new computer system. After logging, they were all dealt with satisfactorily. The main problems were:

3 Leave a clear line space and then import the data from *2009 problems*.

4 Centre align the data on the page.

5 Update *Problems* to save this change.

6 Close and then reopen *Problems*.

7 Now amend the data so that Harry becomes **George**.

8 Widen the display of the data.

9 Add a coloured fill to the heading row.

10 Save as *Problems resolved* and close the file.

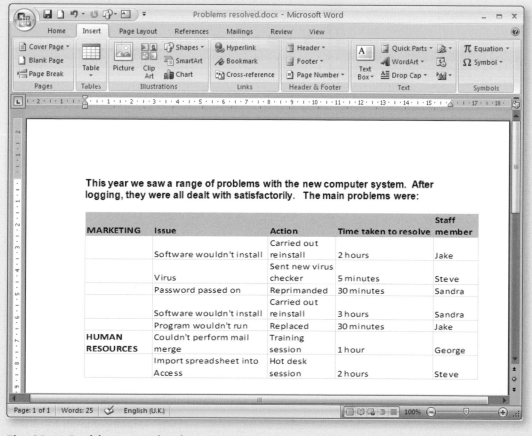

Fig. 22.41 Problems resolved

Importing charts and graphs

There are three different ways to add a chart to a Word document:

- Copy and paste a chart saved elsewhere, for example in an Excel file.
- Insert a data file that contains a chart on a separate sheet.
- Create a chart from scratch. The process is similar to using the wizard to create a chart within Excel.

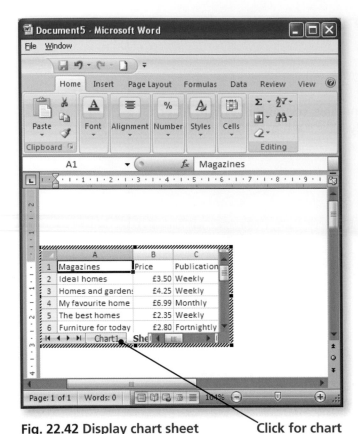

Fig. 22.42 Display chart sheet   **Click for chart**

insert a chart or graph already created

1 Click on the page where you want the chart to appear.

2 Click on the **Insert** tab and then click on the drop-down arrow next to **Object**.

3 Click on **Object** and then click on the **Create from File** tab.

4 Click on **Browse** to locate the data file such as an Excel file containing the graph.

5 Decide whether or not to create a link with the file so that changes are updated automatically. Then click on **OK**.

6 When the data appears on the page, double click for the sheet tabs and scroll bars. You may need to navigate to the chart if it is not on Sheet 1.

7 Use the chart tools to edit or format the graph. (See Unit 023 on spreadsheets for full details).

8 Back in Word, click on the chart and drag out a boundary from a corner if you want to change its size.

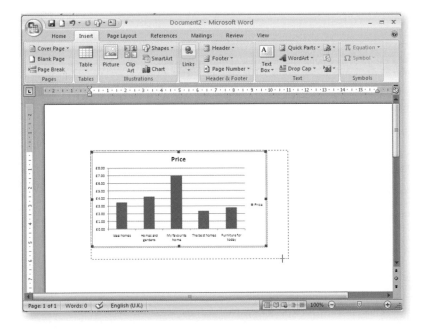

Fig. 22.43 Resize imported chart

insert a new chart directly into Word

1 Click on the **Insert** tab and click on **Chart**.

2 Select the type of chart, for example 2D column, from the range presented.

3 A ready-made chart will appear together with some data already entered in a spreadsheet. This window will be labelled *Chart in Microsoft Office Word*.

4 Start replacing the data with your own in the spreadsheet. The chart will be amended automatically on the page.

5 If you need further columns or rows, drag the lower, right-hand corner of the spreadsheet data area to make room for your entries.

6 When all the data has been added, close the Chart to return to your page.

7 Click on Word's **Chart Tools – Layout** tab to add axes and chart titles.

8 Reopen the spreadsheet data at any time by clicking on the **Edit Data** button.

Chart reflecting changes to data   Reopen spreadsheet data   Chart tools

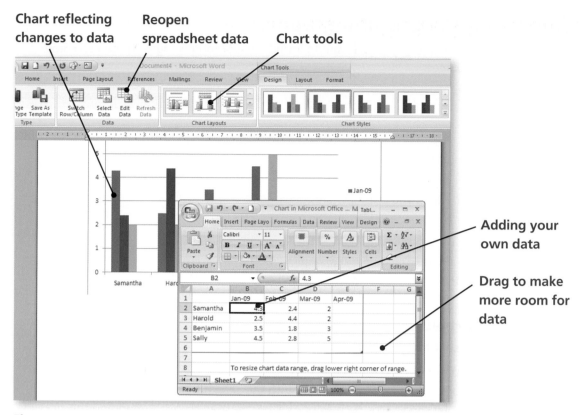

Adding your own data

Drag to make more room for data

Fig. 22.44 Create chart

Check your understanding 10

1 Start a new document.

2 Type the following:

New Soups

Here are the sales figures for our seven new soups. As you can see, Greek Salad was the favourite, closely followed by Mediterranean.

3 Now import the chart from the file *Soups*.

4 Make sure it is large enough for all the data to be seen clearly but does not extend into the page margins.

5 Centre the chart on the page.

6 Add the following text under the chart, making sure you leave at least two clear line spaces:

Many thanks to all the team who made this success possible.

7 Save as *New Soups with Chart* and close the file.

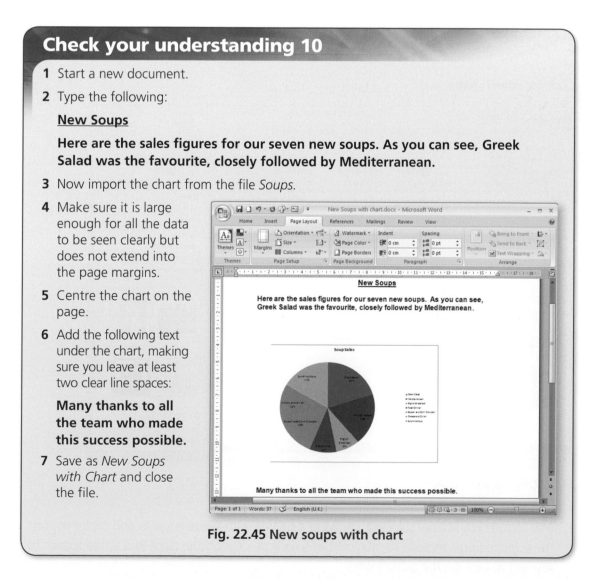

Fig. 22.45 New soups with chart

Importing images and graphics

For many types of document, you will want to include one or more images. These can come from a variety of sources:

- a digital camera
- a scanner
- files previously saved on your computer
- a CD-ROM
- microsoft's Clip Art gallery
- the Internet.

Depending on their origin, you insert pictures in different ways.

Clip Art

Although there is a reasonably wide range of pictures organised into categories available within Word 2007, many people will go online because of its much wider range and variety. Remember that many pictures are copyright and cannot be used for commercial purposes.

insert a Clip Art picture

1 Click on the **Insert** tab.

2 Click on **Clip Art** to open a search box on the right of the screen.

3 Type a brief subject, for example *buildings*, into the search box. (You could restrict your search to particular media files such as photos before searching by amending the ticks in any checkboxes in the **Results should be:** box.)

4 Click on the **Go** button and relevant pictures will appear.

5 Scroll down until you find a picture you like.

6 Click on it once and it will appear on your page.

Click for search pane

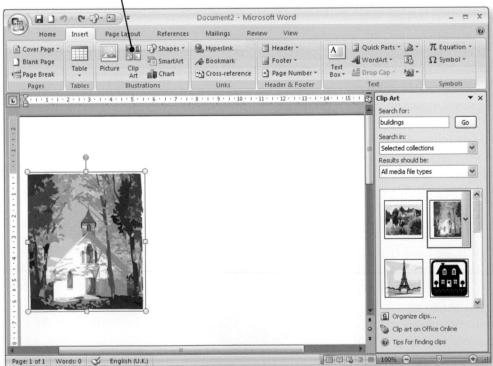

Fig. 22.46 Insert Clip Art

The picture will appear selected, showing a border with white squares and circles round the edge (sizing handles). To take off the selection, click on the page in a white space. To reselect it, click on the picture once.

Saved pictures

If you have saved a picture or have one available on a CD, you can find it and insert it into your document. It can then be edited in the same way as a Clip Art image. The common file types you will come across are:

● **Bitmap** or **raster images** (comprising hundreds of dots of colour called pixels)

> **TIF** – a common format for scanned images
>
> **JPEG** – compressed image files that offer a good range of colours
>
> **GIF** – simpler images
>
> **BITMAP** – large files created using Microsoft Paint

● **Vector images** – these are made up of a collection of curves and lines. They can be scaled down without loss of quality as this is not limited by the number of dots per inch. Common file types include:

> **EPS** (Encapsulated PostScript)
>
> **WMF** (Windows Metafile)
>
> **AI** (Adobe Illustrator)
>
> **CDR** (CorelDraw)

insert a picture from file

1 On the **Insert** tab, click on **Picture**.
2 This opens up the folders structure of your computer.
3 Navigate to the folder containing the picture. It may be in the **My Pictures** folder already created inside **My Documents**, or you may have to go up a level, for example to look in **My Computer** for a CD-ROM in the D: drive.
4 Once you can see the picture you want to insert, click on it in the main window to select it.
5 Click on the **Insert** button.

Select the picture

Click to open Insert Picture dialog box

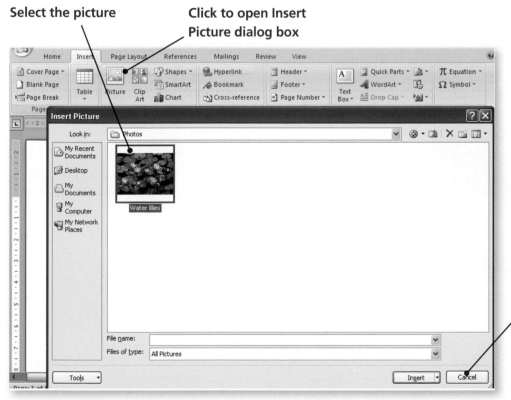

Click to add to document

Fig. 22.47 Insert picture from file

6 When searching for a picture, you can set the computer to display thumbnails or a preview of any image files instead of just their names or details. Do this by selecting an alternative option from the **Views** button.

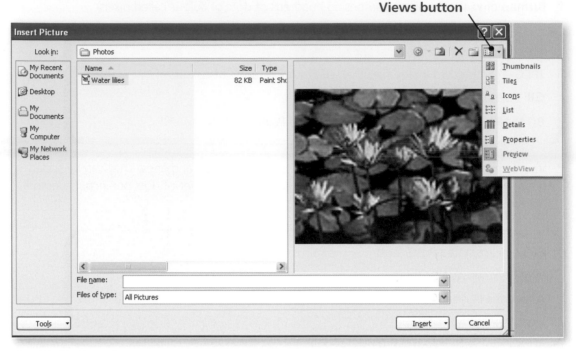

Fig. 22.48 Views button

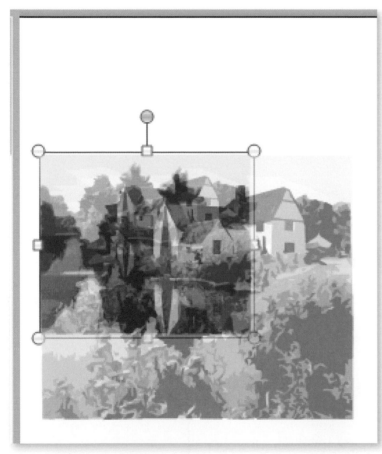

Fig. 22.49 Resize picture

Editing images

Once on the page, you can make a variety of changes to the images.

resize a picture

1 Click on the picture to select it.

2 Position the pointer over a corner sizing handle.

3 When it changes to a two-way arrow, hold down the mouse button and gently drag the pointer inwards or outwards.

4 The pointer becomes a small cross and the newly sized picture will appear as a faded image.

5 Let go when the picture is the correct size.

Note that you should use corner rather than central sizing handles to maintain the picture's proportions.

resize a picture exactly

1 Select the picture and click on the **Picture Tools** tab.
2 Amend the height and/or width measurements in the **Size** boxes.
3 You can also click on the arrow to open the **Size** dialog box. Here, amend measures and, to keep the picture in proportion, make sure there is a tick in the **Lock aspect ratio** box.
4 To increase or decrease the overall size by a percentage, change the setting in the **Scale** boxes.

move a picture

1 Position the pointer over the picture.
2 When the pointer shows a four-way arrow, hold down the mouse button and drag the picture across the screen.
3 The pointer will show a white arrow ending in a small box and you will be able to move the picture in a limited way such as before or after text on the page. Its new position will show as a dotted, vertical bar.
4 You can also use the alignment buttons on the **Home** tab to centre or realign the picture on the page.

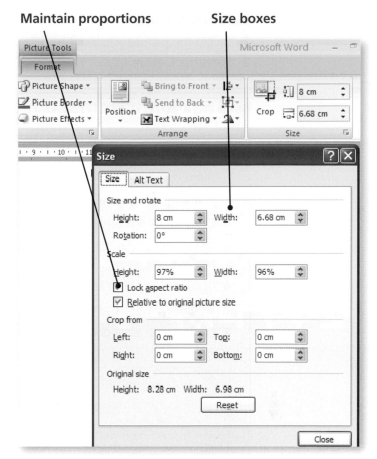

Maintain proportions   **Size boxes**

Fig. 22.50 Exact picture size

Using Text Wrap

It is more flexible to be able to drag a picture round the page, so that it can be positioned accurately. To do this, you need to change its properties. This is done by applying a text wrap. Once applied, the picture can be dragged around with the mouse.

Text wrapping is also used to control how text wraps round an object, for example when an image is positioned alongside text in a document such as within a column.

apply a text wrap

1 Select the picture.
2 On the **Picture Tools** tab, click on **Text Wrapping**.
3 Select an option such as **Tight** or **Square**.
4 You will now find you can drag the picture easily.
5 You can also drag the picture up into a block of text and set how the text flows round it.

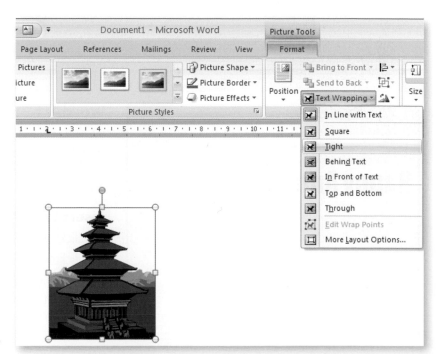

Fig. 22.51 Text wrap picture

delete a picture

1 Select the picture you want to delete.

2 Press the **Delete** key.

Check your understanding 11

1 Open the file *Antiques* on the CD-ROM accompanying this book.

2 Leave a clear line space and then insert a Clip Art picture of any household item after the text **talking about eBay**.

3 Reduce it in size and make sure you keep it in proportion.

4 Centre it on the page.

5 At the end of the document, leave a clear line space and insert the picture *vases.jpg* from the file.

6 Reduce it in size so that the whole document fits on one page.

7 Right align the picture on the page.

8 Save the file as *Antique pictures*.

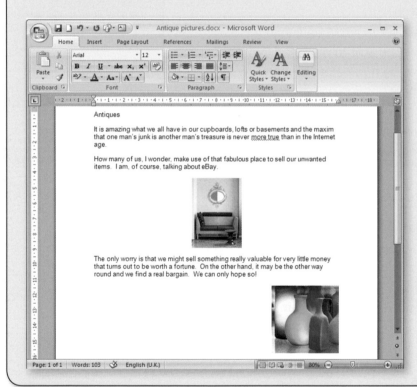

Fig. 22.52 Antique pictures

Fig. 22.53 WordArt insert

WordArt

WordArt is a form of text that that can be shaped, stretched, coloured and moved around a document. It is very useful when you want a title or block of text to stand out.

insert WordArt

1 Click on the **Insert** tab.

2 Click on the **WordArt** button and select a style from the gallery.

3 An editing box will now open and you can type in the text for your WordArt object.

4 Click on **OK** and your text will appear on the page in the style you selected earlier.

5 Now use the various WordArt tools available on the **Format** tab, or right click for different options, to make changes:

   a Scroll through the styles to select a different type of text from the gallery.

   b Add shadows or 3-D effects.

   c Change the main colours of the fill or line.

   d Click on the **Edit shape** button to change the overall shape of the object.

   e Make changes to the text direction or character spacing.

   f Click on the **Size** button to set an exact size.

   g Click on **Arrange** to position the WordArt on the page, rotate it, position it behind or in front of other objects or set a text wrap. This will enable you to drag it into position or send it behind or in front of any text on the page.

6 Select the WordArt and press the **Delete** key to delete it.

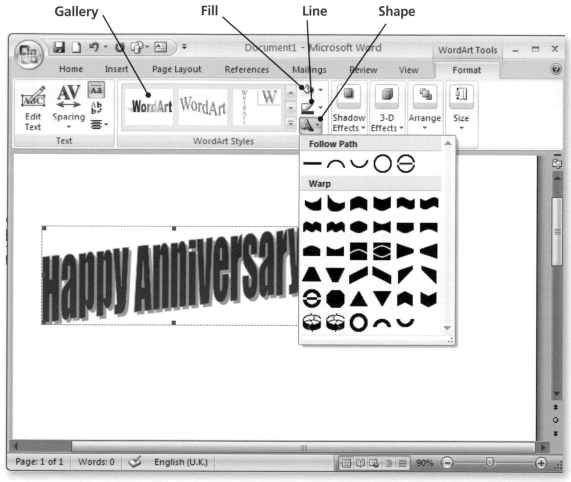

Fig. 22.54 WordArt tools

Shapes

There is a wide range of ready-made lines, shapes and boxes that can be added to documents. They are organised into groups such as **Basic shapes, Arrows, Flowchart shapes** and **Stars** and **Banners**.

Adding and working with shapes was covered at Level 1, but here is a summary of how you can add them to your documents and then change their appearance and position.

add shapes

1 Click on the **Insert** tab.

2 Click on **Shapes**.

3 Select a preferred shape or line from the gallery.

4 Click on the page to add a shape of a pre-set size, or click and drag the mouse to draw out the shape as large or as small as you like.

5 To draw a circle, hold **Ctrl** as you drag out an oval. To draw a square, hold **Ctrl** as you draw a rectangle.

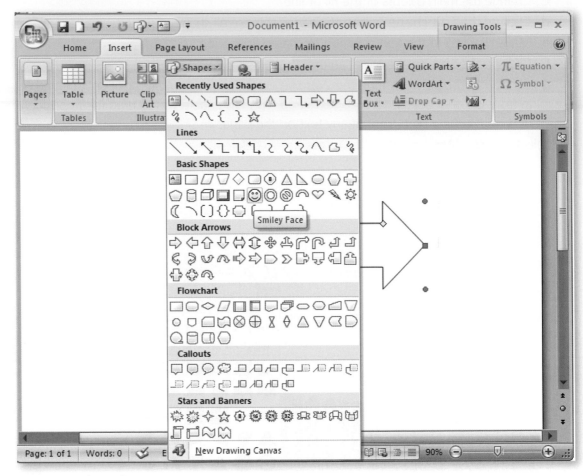

Fig. 22.55 Shape insert

Editing shapes

When the shape appears on the page, formatting tools will be available on the Drawing Tools tab to change the appearance, position and orientation of the shape.

Shapes are often used to contain or border a mixture of images and text and so it is particularly important to be able to arrange all the various objects behind or in front of one another to view them correctly.

edit a shape

1 Make sure the shape is selected. It will show circles and squares around the edge that can be dragged.

2 Click and drag a sizing handle to change the shape size or length of line. The yellow diamond will allow you to alter the actual proportions.

3 Hover the mouse over any of the **Shape Styles** to see the effect before selecting a different visual style.

4 Add a colour fill or line colour, and change the weight or style of line or arrow head from the drop-down menu in the **Line** box.

5 For patterns or textures, right click on the shape and select the **Format** option. On the **Colours and Lines** tab, click on **Fill Effects**.

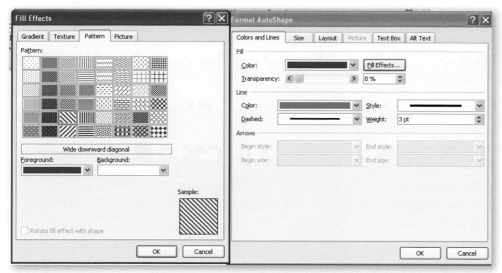

Fig. 22.56 Fill effects

6 Apply a text wrap to set how text wraps round the shape on the page. (Unlike pictures or WordArt, you can always drag a shape to a new position.)

7 Click on the **alignment** button to position the shape in relation to other shapes, margins or the actual page borders.

8 With more than one shape overlaying another on the page, send them backwards or forwards or behind or in front of text.

9 You can also group several shapes into a single entity that can be copied, resized or moved. This is very useful if the new shape is complex and you want to treat it as a single entity to copy or resize. (First select more than one shape by drawing round them all with the mouse or holding **Ctrl** as you click on them in turn.) Click on them individually or ungroup them to edit single shapes that have been grouped. You can then regroup or group them again.

10 Change to a different shape by clicking on the **Change** button.

11 Add text after clicking on the **Edit Text** button. (If the shape inserted is a **Text** box, a cursor will automatically appear inside ready for text entry.)

12 Select the shape and press **Delete** to remove it from the page.

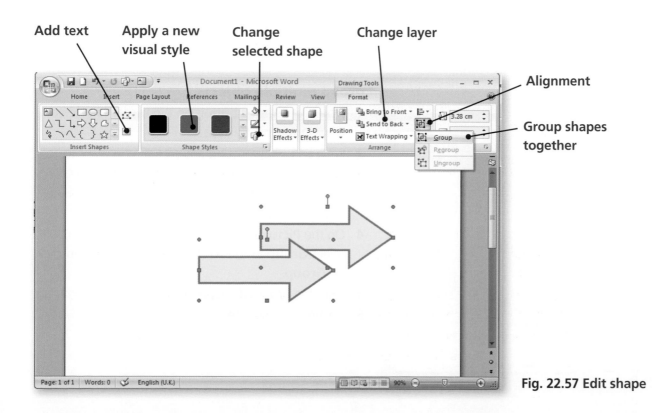

Fig. 22.57 Edit shape

Check your understanding 12

1 Start a new document.

2 Insert a WordArt object reading *For Sale.*

3 Increase it in size so that it fills about a quarter of the page.

4 Position it centrally at the top of the page.

5 Add two circular shapes to the page, one filled with red and one filled with blue, and position them side by side.

6 Group the two shapes and increase their size so that they also fill about a quarter of the page.

7 Position the grouped shapes below the WordArt.

8 Change the WordArt text to read: *Not For Sale* and resize as necessary.

9 Change the colour of the WordArt.

10 Change the red circle to a triangle.

11 Draw a straight line below the shapes.

12 Increase the weight of the line and colour it green.

13 Save as *Shapes.*

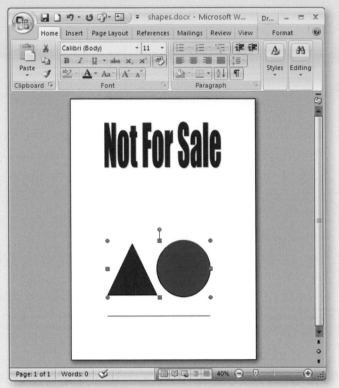

Fig. 22.58 Shapes

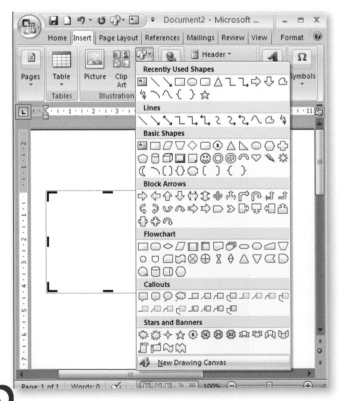

Fig. 22.59 Canvas

Grouping pictures and shapes together

To do this, you need to use the **Drawing Canvas** – a special area used for working with images and shapes.

group multiple objects

1 From the **Shapes** button on the **Insert** tab, select **New Drawing Canvas**.

2 Set a text wrap around any picture you want to group and then cut and paste it onto the canvas.

3 Repeat this with other pictures and the shapes you want to group it with.

4 On the **Picture/Drawing Tools – Format** tab, select **Group** in the **Arrange** section and then click on **Group**.

5 Drag the grouped shape off the canvas.

6 Delete the unwanted canvas.

Sections

When compiling long documents that incorporate imported material such as charts or tables of data, it may be necessary to retain the integrity of the additions in terms of their orientation, margins, headers and footers and so on. To do this, you need to divide the document into sections. Each section can then have its own layout.

create sections

1 Click in place for the next section.

2 On the **Page Layout** tab, click on **Breaks**.

3 Select **Section Breaks**.

4 You can select the option to start the next section on a new page or for it to continue from this point.

5 You can now make layout changes to this section of the document only.

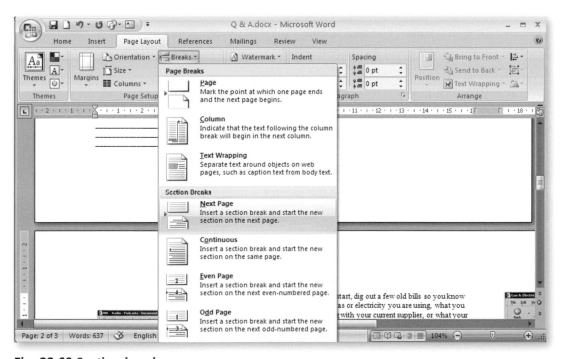

Fig. 22.60 Section break

Setting text in columns

There are three different ways you can line up blocks of text across the page:

- using tabs
- creating tables
- formatting text in columns.

Using tabs

Each time you press the **Tab** key (labelled with two arrows, next to the letter Q) it moves the cursor across the page. The normal distance it moves is in jumps of 1.27 cm (0.5 in). If you fix the position it moves to (known as the **Tab stop position**), you can use this facility to help you set out information in a document in neat columns. You do this by placing black symbols along the ruler exactly where you want the cursor to move to for each column.

As well as the tab stop position, you can also set the *way* in which columns line up (**Tab alignment**):

- by their initial characters (**Left Tab**)
- by their last characters (**Right Tab**)
- centred on the tab position (**Centre Tab**)
- with decimal places in a column of figures lining up exactly (**Decimal Tab**).

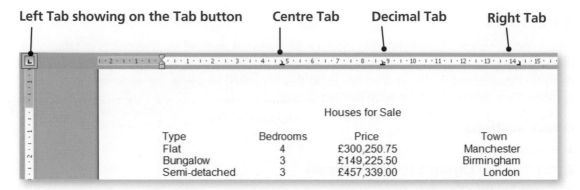

Fig. 22.61 Columns with tabs

fix tab stops using the ruler

1 Make sure the horizontal ruler is showing.

2 Start by typing any column headings, spacing these by eye. It will be easier to get their position correct after typing all the text if they are *not* set using the **Tab** key.

3 Press **Enter** to position the cursor on the left margin.

4 Set the position for the first column. If you do not want a Left Tab, first click on the **Tab** button above the left scroll bar until it shows the correct type of tab symbol, for example **Right** or **Centre**, then click on the ruler where you want the cursor to move to. The chosen symbol will appear.

5 Repeat the process until all the tab stops are on the ruler.

6 If necessary, first click on the **Tab** button again for a different alignment.

7 If you make a mistake, rest the pointer on the tab stop and then gently drag it up or down away from the ruler. When you let go of the mouse the symbol should disappear. You can also drag a tab stop to a different position along the ruler.

8 To type your document, either type the first entry so that the first column lines up on the left margin or press the **Tab** key and start typing where the cursor moves to. Press the **Tab** key again to move the cursor to the next column and repeat across the page.

9 When you have completed the first row, press **Enter** to move to the next line and start typing the next row of entries.

10 If you find you need a wider column, select **all** the text typed using tabs. Now drag the position for the incorrect tab along the ruler. A dotted line will show its progress and the column position will be readjusted.

use the menu

1 Type the headings and then press **Enter** to move to the first line for the column entries.

2 Type the first entry and then press the **Tab** key once to move to the next column. Repeat across the page. Your text will appear very squashed up but will be sorted out later.

3 Repeat for the remaining entries, pressing **Enter** each time to move to the next row. Then highlight all the column entries set with tabs.

Type	Bedrooms	Price	Town
Flat	4	£3550000	London
Maisonette	3	£127,580	Manchester
House	5	£467890	Birmingham
Semi-detached	4	£334560	Exeter

Table 22.2 Single tab between entries

4 On the **Home** tab, click on the arrow to open the **Paragraph** dialog box and click on the **Tabs** button.

5 In the **Tabs** dialog box, click in the **Tab stop** position box and enter the measure (figures only, no cm) for the first column, for example Bedroom entries would be positioned in a column lined up at 4 cm along the ruler by entering **4**. Click for **Centre**, **Right** or **Left** style of tab and then click on the **Set** button.

6 Click in the box again (you will have to type over the entry that appears) and enter the position and tab alignment style for the next column. Repeat until all tab stop positions have been set.

7 Return to your document and check that the columns look right. If not, return to the dialog box and change any measurements or tab styles.

8 To remove unwanted tab stops, select them in the box and click on the **Clear** button.

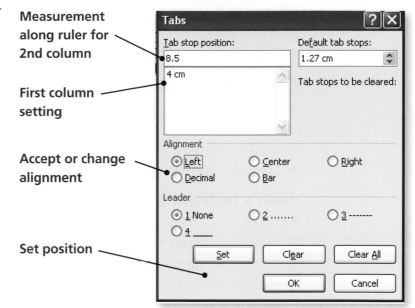

Fig. 22.62 Set tabs with menu

1 Start a new document.

2 Enter the title *Animals for Food*

3 Save the file with the same name.

4 Centre the title and apply uppercase.

5 Now create the table set out below. Enter the column headings first and then use tabs.

6 Set the columns as follows:

   • Type – on the left margin
   • Meat – Left tab at 3 cm
   • Price per kilo – Decimal tab at 7 cm
   • Varieties – Right tab at 13 cm

Type	Meat	Price per kilo	Varieties
Pig	Pork	£8.45	Bacon, Ham
Cow	Beef	£12.75	Veal
Lamb	Mutton	£9	Lamb

7 Format the column headings in bold and italic and make sure they are positioned correctly over the column entries.

8 Finally, move the Price per kilo column to 8.5 cm and adjust the headings so that they are still over the column entries.

9 Save and close the file.

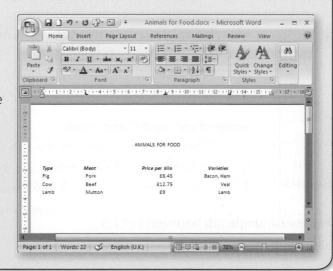

Fig. 22.63 Animals for food

Working with tables

Tables are a series of rows and columns of cells. Each cell contains discrete data that can be aligned, formatted and edited separately. Single rows and columns or the whole table can have visible gridlines, you can add emphasis in the form of enhanced lines and shading or, by removing the gridlines, the data can be displayed in simple columns.

Once the table has been created and you want to start entering the data, move from cell to cell using the **Tab** key or mouse.

Creating a table

There is an option available to draw your own table if it has a complex shape, but normally you use one of two methods to set up the correct number of columns and rows.

- using the mouse
- using the table menu.

insert a table using the mouse

1 Click on the page where you want the table to appear.
2 Click on the **Insert** tab.
3 Click on the **Table** button.
4 Drag the mouse across the rows and columns of cells to create your preferred size of table. You will see it taking shape on the page.
5 Let go of the mouse and you will return to your page.

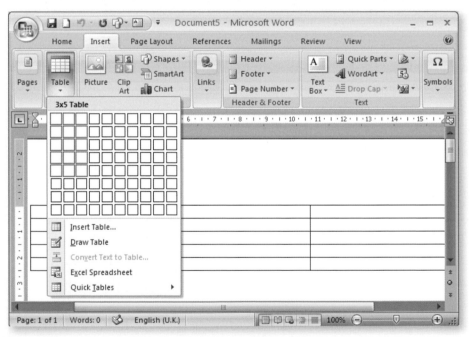

Fig. 22.64
Insert table

Fig. 22.65 Insert table menu

insert a table using the table menu

1 Click on the **Insert Table** option.
2 Enter the required number of columns and rows.
3 Click on **OK** and the table will appear.

amend columns and rows

1 Click in the last cell and press the **Tab** key to add a new row of cells.
2 Right click any cell, select **Insert** and select the appropriate option such as a new column to the right or a row above.
3 Right click on a cell, select **Delete cells** and then select an option such as an entire row or column.

4   You can also click on the **Layout** tab under **Table Tools** and click on an **Insert** or **Delete** option.

5   If you click on the small square in the top left-hand corner of the table and select the whole table, pressing the **Delete** key will delete cell contents rather than the actual table.

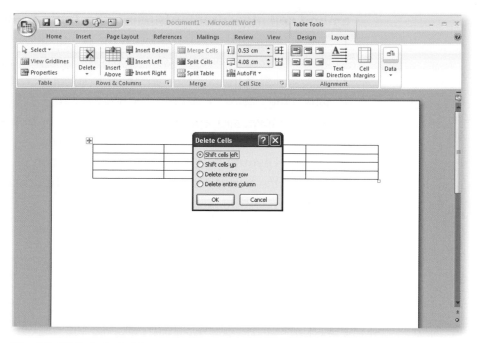

Fig. 22.66 Delete table

Column widths and height

The default size for each table cell depends on the number of columns in the table and the entries you have made. If the default measurements do not allow you to display your data appropriately, you can amend individual cell sizes or that of an entire column or row.

amend measurements manually

1   To change column width, move the mouse up to the ruler and hover over the blue marker showing the right-hand column boundary.

2   When it displays a black two-way arrow, click and hold down the mouse button.

3   Gently drag the border to the right to increase the width of the column.

4   You can also drag a column border with the mouse from within the table.

5   If you double click on the cell boundary within a table, it will set the width to the longest entry.

6   For row height, follow the same process but drag the boundary vertically up or down.

7   If you double click on the ruler, you will open the **Table Properties** box and can make other changes such as setting measurements for each row or column exactly.

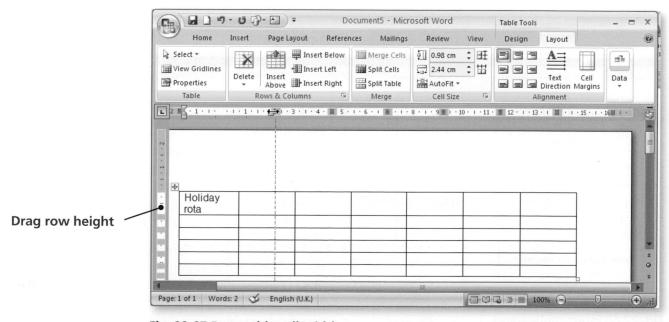

Drag row height

Fig. 22.67 Drag table cell width

amend cell measurements using the Table Tools

1 Select the target column by moving the pointer above it until it shows a down-facing black arrow. Click and the column cells will turn blue.

2 Click and drag the pointer if you want to select more than one column to adjust at the same time.

3 To select rows, click and drag the cell contents or click and drag the pointer in the left margin when it shows a right-facing white arrow.

4 Click on **Layout** under **Table Tools** on the ribbon.

5 Click on the arrows in the **cell width or height** box to increase or decrease the measure, or enter your own figures and press **Enter**.

Row height   Column width measure

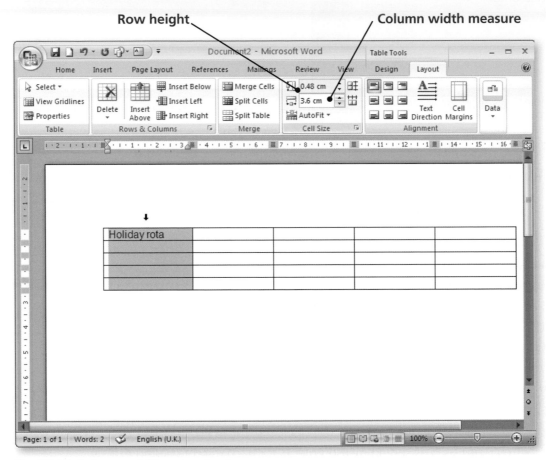

Fig. 22.68 Column width using tools

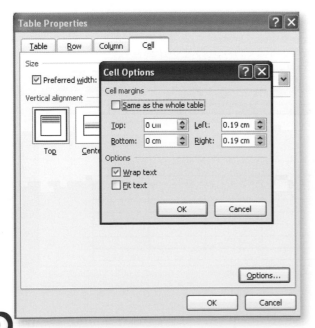

Cell margins

Where you want the text in each cell set in further from the cell borders, you can increase the size of the cell margin.

change cell margins

1 Select the cells.

2 On the **Table Tools – Layout** tab click on the small arrow in the **Cell Size** group to open the **Table Properties** dialog box.

3 Click on the **Cell** tab and then click on the **Options** button.

4 Take off the tick in the **Same as whole table** box and change measurements in the appropriate margin boxes.

Fig. 22.69 Cell margins

Cell alignment

Entries in cells can be aligned vertically or horizontally if you want to change how the data is displayed.

Horizontal

Left aligned entry	Centre aligned entry	Right aligned entry

Vertical

Top alignment		
	Centre alignment	
		Bottom alignment

You can also combine the alignments, for example **Bottom Left** or **Top Right**, so that there are nine different alignments that can be set.

set cell alignment

1 Select the target cell(s).

2 Right click and select **Alignment**.

3 Click on the appropriate button.

4 Or find these on the **Layout** tab under **Table Tools**.

5 You can also set horizontal alignment using the **text alignment** buttons on the **Home** tab.

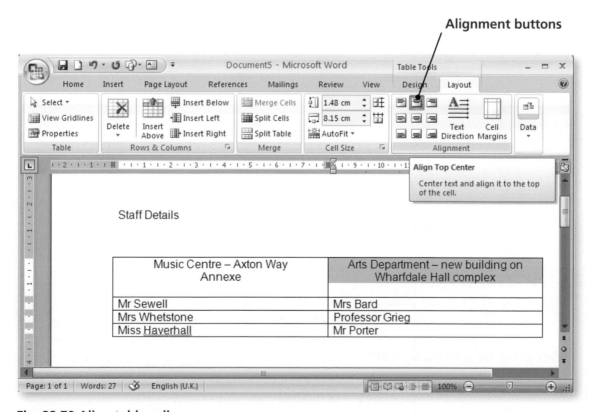

Fig. 22.70 Align table cell

Check your understanding 14

1 Start a new document and save as *Beads*.

2 Enter the title **Beads**. Set the font size to 20 and centre the heading on the page.

3 Create a table that has four columns and five rows.

4 Enter the following four headings:

Type of bead   Main colour   Size   Cost per pack of 10 beads

5 Now add the following data:

Type of bead	Main colour	Size	Cost per pack of 10 beads
Lampwork	Multi	8mm	£5
Rocaille	Gold	1mm	£3.50
Bugle	Silver	4mm	£4
Bicone	Blue	10mm	£8.25

6 Set the heading row height to 1.5 cm.

7 Increase the font size for the heading text to 16 and the main entries to 14.

8 Align the column headings centrally at the top of the row.

9 Change the width of the final column if necessary so that the heading is on three lines.

10 Realign the first three headings so they are centred vertically as well as horizontally.

11 Centre the table on the page. Either use the **alignment** buttons on the **Home** tab or move the table by dragging the move handle in the top left-hand corner showing four arrows.

12 Save the file to update these changes.

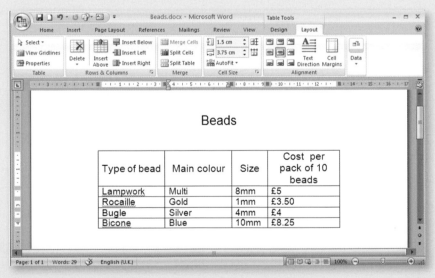

Fig. 22.71 Beads table

Merging or splitting table cells

To be able to centre a heading above several columns in a table, you will need to merge the cells. You can also split a cell if you want two or more entries in the original space.

merge cells

1 Select the cells you want to merge.

2 On the **Table Tools – Layout** tab, click on the **Merge Cells** button.

3 The selected cells will now become a single cell.

4 Format the cell contents in the normal way.

Two cells already merged   Merge button   Split cells   Selected cells

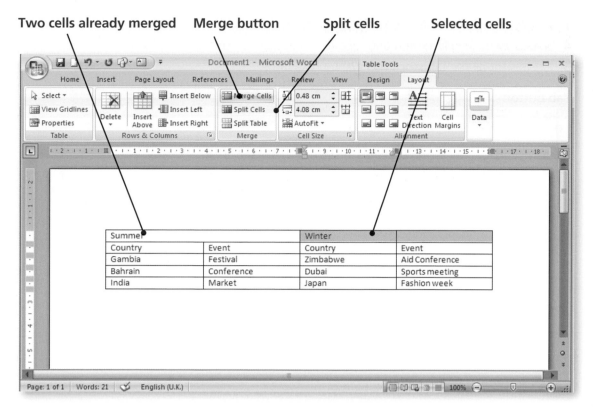

Fig. 22.72 Merge cells

split cells

1 Select the cell(s).
2 Click on the **Split Cells** button on the **Table Tools – Layout** tab.
3 Select the number of columns and/or rows to create.
4 Click on **OK**.

Using tabs in a table

In some cases, you may want to indent table entries within a cell. You can move across a cell in jumps of 1.27 cm (0.5 in) when using the **Tab** key, but to indent entries at an exact point on the ruler, or to set the way they line up, you will need to set tab stops at the position you want the cursor to move to.

As you will move from cell to cell if you press the **Tab** key on its own, you need to hold down **Ctrl** to move to a tab stop set *within* a cell unless it is a Decimal tab.

set tabs within a table

1 Click on the cell in which you want the tab stop. For the same tab stop for an entire column, select all the column cells first.
2 Click on the **Tab** button to the left of the ruler until the correct style of tab is showing – cycling through left, centre, right and decimal.

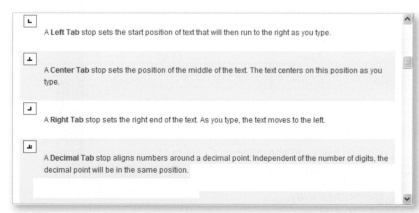

Fig. 22.73 Tab styles

3 Click on the ruler for the tab stop position.

4 Select the next cell/column and repeat.

5 Do this for all the tab stops you want in the table.

Tab button

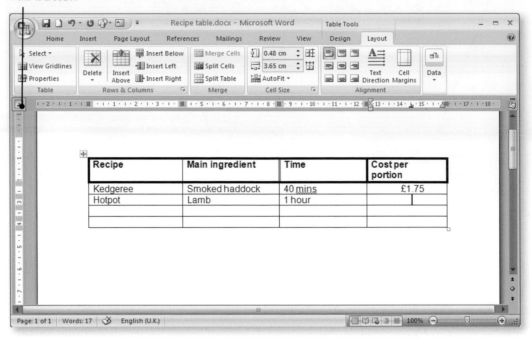

Fig. 22.74 Tab set

use tabs in a table

1 Click in the first cell and enter any data as normal.

2 Move across the table and to the cell containing the tab stops by pressing the **Tab** key.

3 For decimal tabs, type in the figure. The cursor automatically moves to the tab stop position.

4 For other types of tab, hold **Ctrl** and press **Tab** to move the cursor to the tab stop position within the cell.

5 Continue to enter data into the table in the normal way.

Left tab stop in cell

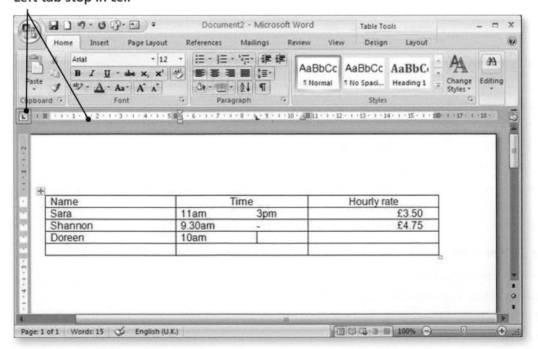

Fig. 22.75 Using tabs in a table

Check your understanding 15

1 Start a new document and create a table with four rows and three columns.

2 You will be entering the data as shown below.

3 You want to indent the names. Set a **Left** tab in the first column 1 cm in from the cell boundary.

4 Set a **Left** tab in the second column 2 cm in from the cell boundary to indent the course data.

5 Enter column headings without using tabs, and centre the headings in the first row.

6 In the first cell, move to the tab stop and enter the text **Sally**.

7 Move to the next cell. Type the two lines of text. (To add entries on a new line within the same cell, press **Enter**.)

8 Move to line 3 and then move the cursor to the tab stop position to enter **Advanced Word**.

9 Move to the next line and type **Web pages** at the tab stop position.

10 Use the **Tab** key to move to the next cell to enter the grade.

11 Enter the rest of the data, lining up entries as shown.

12 Save as *Training table.*

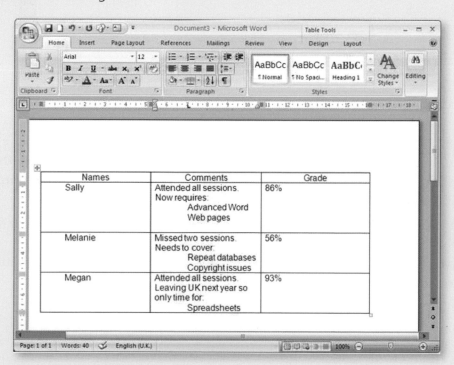

Fig. 22.76 Training table with tabs

Borders and shading

To add more emphasis to a table, you can change the width and style of cell border and colour and shade both the lines and the cell backgrounds. Alternatively, you can remove borders to display just the data in columns on the page.

change borders

1 Select the target cells.

2 Click on the **Table Tools – Design** tab.

3 Click on the top **Line Style** button drop-down arrow to change from a single line to double lines or dashes, and the **Line Weight** arrow to select a line thickness.

4   Then click on the **Borders** button and select an option such as **Outside Border** to apply the border.

5   If you change line style or weight, remember to click on the **Borders** button again to apply the new settings.

6   Remove borders by selecting **No Border** from the Borders or Line Style menus.

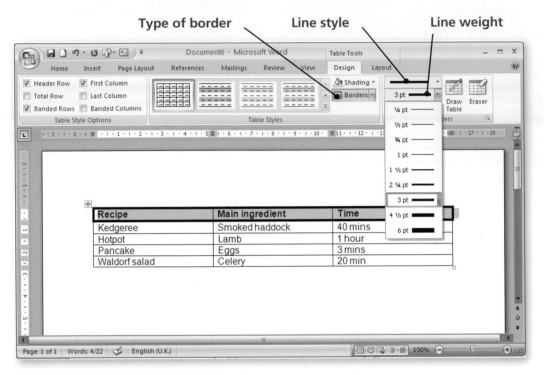

Fig. 22.77 Border buttons

Or

7   Right click in the table and select **Borders and Shading** to open the dialog box. This option is also available from the small arrow in the bottom corner of the **Draw Borders** box on the ribbon.

8   You can now select border style, weight and type from the various boxes in the same way that you did for border text.

9   To remove a partial border, such as a horizontal top border, click on the correct button in the **Preview** section.

10   You can also change the colour of the border line.

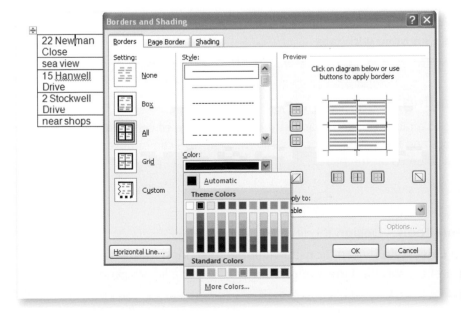

Fig. 22.78 Borders dialog box

shade cells

1 Select the target cells.

2 On the **Table Tools – Design** tab, click on the drop-down arrow in the **Shading** box to select from a range of different colours.

3 You can preview the effect by resting your mouse on the selected colour.

4 If you choose from the themes rather than standard colours, this may also change the appearance of the fonts you are using.

5 Click on **More Colors** for a wider palette.

6 Remove any shading by selecting **No Color**.

7 You can also apply a pre-set table style offering different borders and shading by clicking on one of the examples on the ribbon.

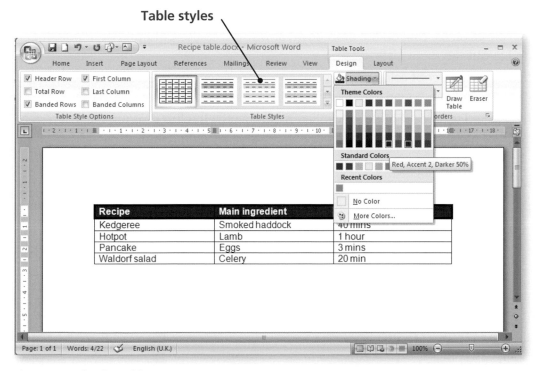

Fig. 22.79 Shade table

Check your understanding 16

1 Reopen *Beads*.

2 Add a thick, dark border to the outside of the entire table.

3 Shade the heading row mid-blue.

4 Shade the bead names pink.

5 Save as *Shaded beads*.

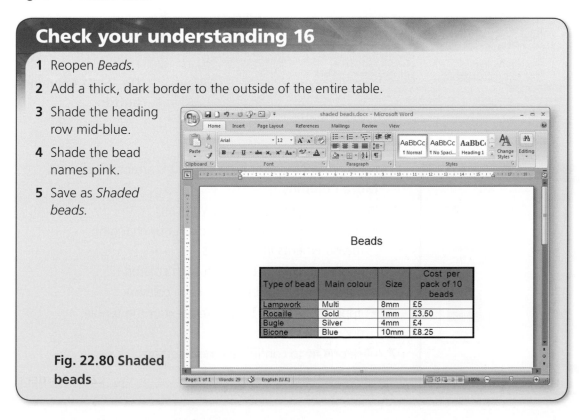

Fig. 22.80 Shaded beads

Columns

The simplest way to display text in two or three columns is to use the **Columns** facility. This divides up the paper so that when your typing reaches the bottom of the page it continues into a second column rather than moving onto a new page.

As you may want the columns of text to balance, or if there is not enough text to actually reach the bottom of the first paragraph, you can start the second or third columns at a set point by inserting a column break.

Make sure that any headings you may want are typed first, before setting columns, or they will be placed over the first column.

set columns

1 Select the text or set columns before typing by clicking on the page where you want the first column to start.

2 For simple columns, click on the drop-down arrow next to the **Columns** button on the **Page Layout** tab.

3 Click on the number of columns you want and you will see any selected text line up on the left. The column boundaries will show as blue marks on the ruler.

4 If the selected text needs more space than is provided by the first column, the rest of the text will be moved into the second or even third columns automatically.

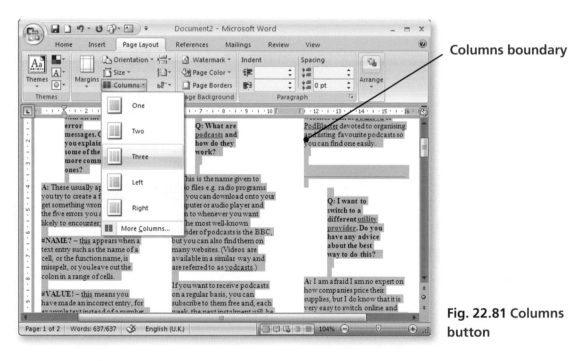

Columns boundary

Fig. 22.81 Columns button

Adjust spacing   Add vertical line

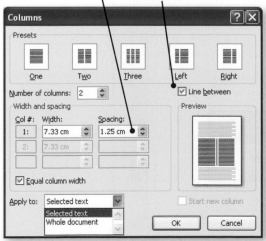

5 For more control over how the text is arranged, click on **More Columns**.

6 In the dialog box that opens you can:
   - set exact widths for each column by changing measurements in the boxes
   - add a vertical line between each column
   - amend the space between each column
   - apply the settings to selected text or the whole document.

7 Click on **OK** to confirm the new settings.

Fig. 22.82 Column dialog box

continue typing below columns

1 If you have just typed text that is set in columns, press **Enter** to move to a new line.

2 Open the **Columns** dialog box.

3 Click on the **One** Presets (as normal text is actually a single, page wide column).

4 Click in the **Apply to:** box and select the option **This point forward**.

5 Click on **OK**.

6 You will now find the cursor has moved and as you type your entry it will extend across the full page.

insert column breaks

1 Click in front of the word you want to start in a new column.

2 On the **Page Layout** tab, click on **Breaks**.

3 Select **Column** from those shown.

4 Click on **OK**.

You may find that the next text entry has actually moved onto a new page. This is because a section break has been set at the end of the last column. To remove the extra page without affecting your column settings:

1 View the document in **Draft View** (available from the **View** tab).

Fig. 22.83 Column break

2 Click on the **paragraph mark** or 'Section Break' dotted line.

3 Press the **Delete** key.

Check your understanding 17

1 Open the file *Shakespeare* on the CD-ROM accompanying this book.

2 Set the text in two columns.

3 Make sure the text **The insolence of office** starts at the top of the second column.

4 Underneath the quotation, type the following as a normal line of text *not* set in columns: **Probably one of the best-known lines in English literature, this is taken from William Shakespeare's tragedy, *Hamlet*.**

5 Save as *Hamlet* and close the file.

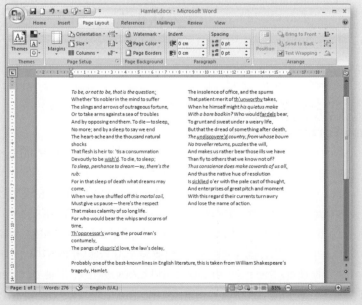

Fig. 22.84 Hamlet

Page layout

At Level 1, you learned how to enter, format and edit simple documents. You should be able to change font, font size and font colour and apply a text alignment such as centred or justified. These options are available on the **Home** tab, from the relevant dialog box and also from a floating toolbar offering a limited range of tools that appears in Word 2007 whenever text is selected.

Open dialog box   **Floating toolbar**

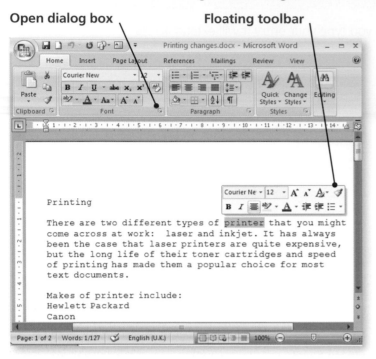

Fig. 22.85 Floating formatting toolbar

You should also be able to delete or insert entries and move or copy text or objects. Here is a reminder of the steps to take when moving or copying.

move or copy text

1 Select the word(s) or block of text.
2 Click on the **Cut** icon to move or the **Copy** item to copy. These are available by right clicking or on the **Home** tab in the Clipboard area.
3 Click on the page or open the document where the text is to appear.
4 Click on the **Paste** icon.
5 If necessary, adjust any spacing.

At Level 2, the emphasis is more on using some of the advanced features of the software to create and edit long documents and make sure they are accurately and professionally laid out.

The main aspects of document layout that you need to know how to change include line or paragraph spacing, margins, page orientation and page and other breaks in the text. (To make changes to a whole document, select it quickly by holding down **Ctrl** and pressing the key **A**.)

House style

Most organisations have a house style by which their documentation can be recognised. Usually this involves using a specific type of font or range of font colours, specified page margins and a certain type of border for images or logos. They may also prefer serif fonts such as Times New Roman or sans serif fonts such as **Calibri**. For example:

- all headings might have to be Arial, font size 16, red
- all subheadings might have to be Arial, font size 14, black

When asked to edit a document and apply the house style, it can take time to change the formatting for each heading and subheading. A quick way to do this is to use the Format Painter utility to literally *paint* a format onto selected text.

use Format Painter

1 Select any part of the text that already has the correct formatting applied, or format some text first and then select this.
2 Click on the **Format Painter** button on the **Home** tab. (To repeat the formatting throughout a long document, double click on the button to keep it turned on.)
3 Use the scroll bars if necessary to display the first entry where the new format needs to be applied. (If you click on the page you will cancel the process or format the wrong text.)
4 Gently drag the pointer along the block of text. The pointer will display a small brush.
5 When you let go of the mouse, the text will have taken on the copied formatting.
6 If necessary, continue throughout the document and then click on the **Format Painter** button to turn it off.

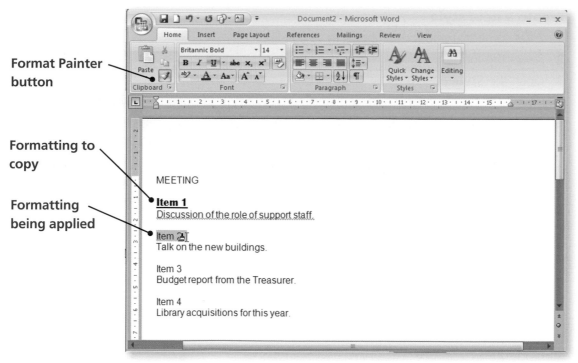

Format Painter button

Formatting to copy

Formatting being applied

Fig. 22.86 Format painter

Styles

A more professional way to format headings and subheadings is to use styles. These are a combination of font and paragraph settings that are named and can be applied directly to selected text. You can either use a ready-made style or create and save your own.

In Word 2007, whenever the mouse is resting on a named style, selected text will change so that you can preview the effect.

see the styles already available in Word 2007

- Scroll through the range of styles showing in the **Styles** section of the **Home** tab. For example, **Heading 1** will apply a Cambria font size 14 blue.
- Click on the **Change Styles** button marked AA to open a window offering a range of style types, fonts and colours.
- Click on the small arrow in the **Styles** section to display a wide range of different styles in a separate **Styles** pane.

apply a style

1 Select the text.
2 Scroll through the options and select a style.
3 Apply repeatedly or use **Format Painter** to copy the style to other headings at the same level.

Styles available

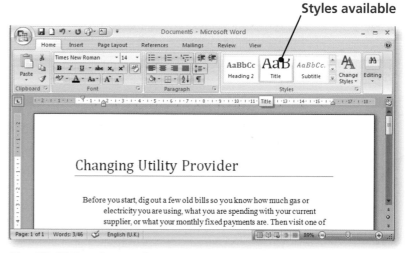

Fig. 22.87 Preview style

Open pane

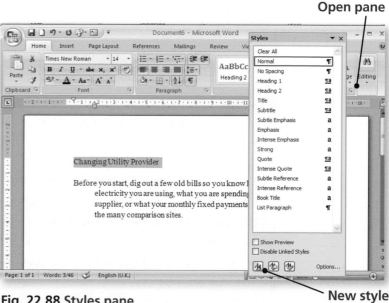

New style

Fig. 22.88 Styles pane

create a new style

1 Enter some text and format it exactly as you like.

2 Select the formatted text.

3 Open the **Style** pane.

4 Click on the bottom left-hand icon labelled **New Style** to open the dialog box.

5 Check the various settings: create your own name for the style, change any alignment, colour or font types and apply the style settings to characters, paragraphs or both.

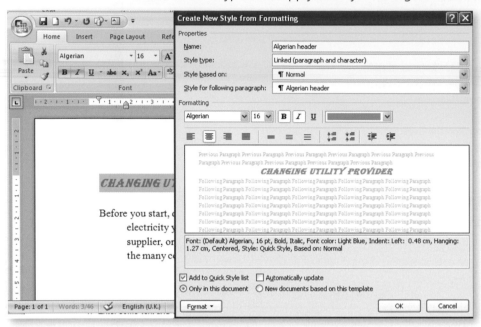

6 Check whether you want the style available only in the current document or in all future documents.

7 Click to create the style and make it available to use. It will have been added to the style pane and will show in the gallery.

8 To reopen the box to make changes to your new style, right click on the example in the gallery and click on **Modify**.

Fig. 22.89 New style

Check your understanding 18

1 Open *Shakespeare* on the CD-ROM accompanying this book.

2 Add a heading *Famous Quotations*.

3 Leave two lines and add a subheading *Hamlet*.

4 Make sure this is at least one line above the start of the text.

5 Save the file as *headings*.

6 Apply the style **Title** from the gallery to the heading *Famous Quotations*.

7 Format the subheading *Hamlet* as follows: Comic sans, italic, size 17, red font colour.

8 Create a new style called *Shakespeare* based on this formatting.

9 At the end of the document, type the play name *Midsummer Night's Dream*.

10 Apply the style *Shakespeare* to this heading.

11 Save and close the file.

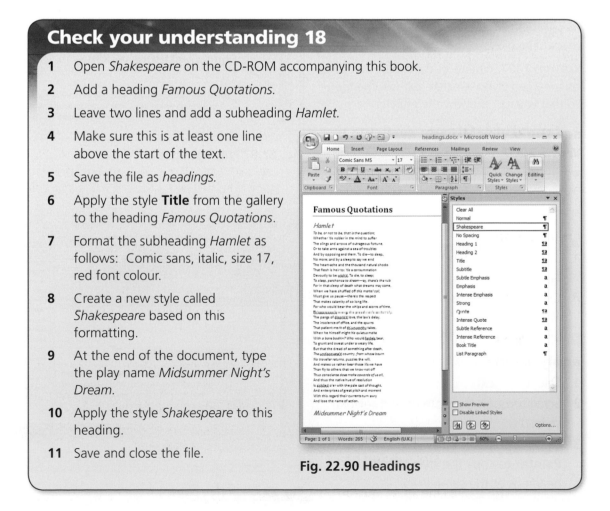

Fig. 22.90 Headings

Paragraph and line spacing

As well as adjusting the spacing between individual lines in a paragraph, you can set the amount of white space left before or after whole paragraphs or selected headings or subheadings.

set special paragraphs

1 Open the **Paragraph** dialog box.
2 For text to be indented in the first line but positioned on the left margin for the rest of the paragraph, click in the **Special** box and select **First line indent**.
3 For the first line to be aligned on the left margin but the rest of the paragraph indented, select **Hanging indent**.
4 In both cases, change the position for the indent by entering an exact measure in the **By:** box, or drag the indent marker along the ruler.

Hanging indent measure changed

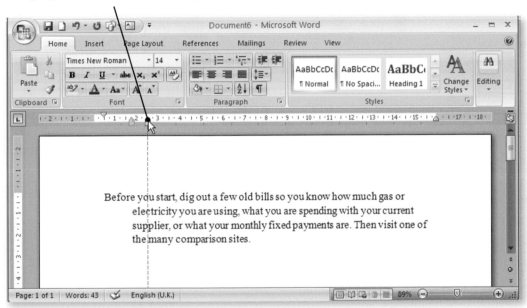

Fig. 22.91 Change hanging indent

Line spacing

The default line spacing is single – each line follows on from the one above with no spaces. This means that every time you press **Enter,** the cursor should move to the line below, whether it is at the end of a sentence within a paragraph or it is after a heading or subheading. To change this setting so there is white space between lines, you can set double, 1.5 or other line spacing.

change line spacing

1 Select the paragraphs or whole document.
2 Use shortcuts to set:
    • double line spacing (**Ctrl** plus **2**)
    • 1.5 line spacing (**Ctrl** plus **5**)
    • single line spacing (**Ctrl** plus **1**).
    Or
3 Select options from the **Line spacing** button on the **Home** tab.

Open dialog box

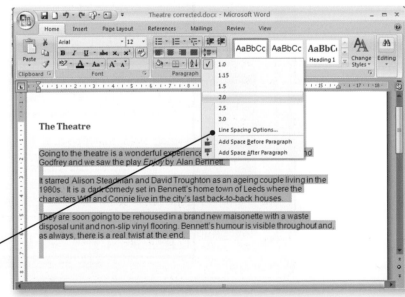

Fig. 22.92 Line space button

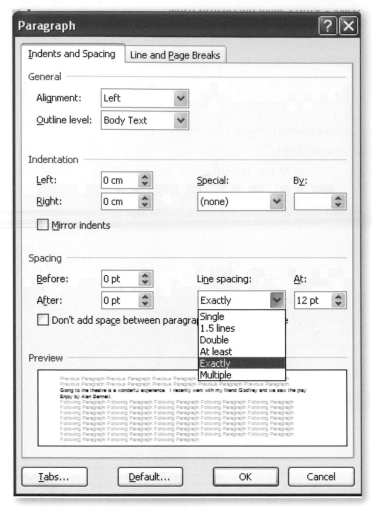

Fig. 22.93 Line space dialog box

Or

4 Right click or click on **Line Spacing** options to open the **Paragraph** dialog box.

5 Here you can choose from a range of line spacings or even set an exact measure. To do this, select an option such as **Exactly** from the line spacing drop-down list and enter the measure in the **At** box.

add line spacing before or after paragraphs

1 Select the heading or paragraph(s) to change.

2 Select a simple **Add Space** option from the **Line Spacing** button.

3 For a specific measure, open the **Paragraph** dialog box from the **Home** tab or by right clicking and selecting **Paragraph**.

4 In the **Spacing** section, click on the arrows or enter exact measures in the **Before:** or **After:** box.

5 The default measurement unit is points, but you can change to inches or centimetres if you add ' or cm in the box. Next time you look, these will have been converted back to the appropriate number of points.

6 Click on **OK**.

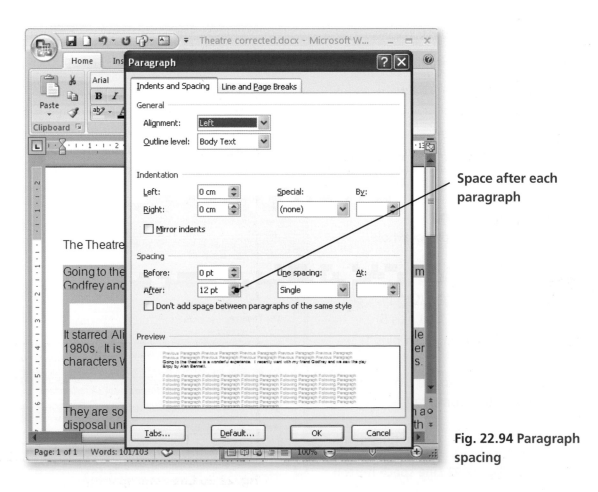

Space after each paragraph

Fig. 22.94 Paragraph spacing

Widows and orphans

If the last line from a previous paragraph starts on a new page, this is known as a **widow**, and if the first line of a new paragraph appears on its own at the bottom of the page, this is known as an **orphan**. These should be prevented from occurring in business documents. By setting the correct paragraph formatting option, single lines will be joined automatically to their appropriate paragraphs.

Note that it is normal to follow a heading or subheading with at least two lines of related text on the same page.

prevent widows and orphans

1 Select the paragraph or whole document.
2 Open the **Paragraph** dialog box.
3 Click on the **Line and Page Breaks** tab.
4 Check that there is a tick in the **Widow/ Orphan control** box.

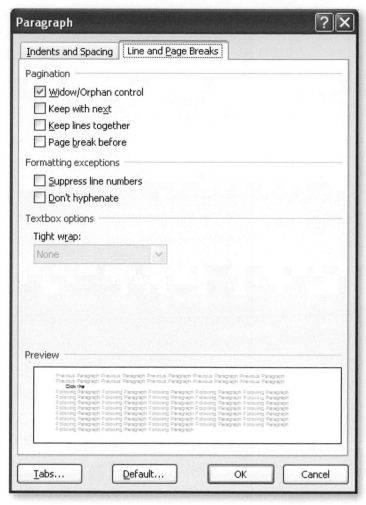

Fig. 22.95 Widows

Horizontal spacing

As well as setting font size and line spacing, you may want your text to take up more or less space along a line. You can change the space between individual letters by altering the measure in the spacing box.

change character ho izontal spacing

1 Select the text
2 Open the **Font** dialog box.
3 Click on the **Character Spacing** tab.
4 In the **Spacing** box, select **Expanded**.
5 Increase the measure in the **By:** box.
6 Check the preview.
7 Click on **OK** to confirm the new setting.

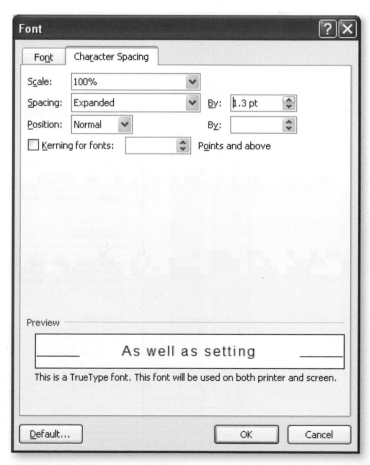

Fig. 22.96 Character spacing

Check your understanding 19

1 Open the text file *York* on the CD-ROM accompanying this book.

2 Format the entire document as follows:

- Font – Times New Roman
- Font size – 11
- Alignment – fully justified
- Emphasis – none
- Line spacing – single

3 Save as a Word 2007 document with the name *York formatted*.

4 Set the spacing after each paragraph to 6 pt.

5 Centre the heading and make it bold, font size 14.

6 Format the subheading *Location* as follows:

- Underlined
- Italic
- Font size 12.

7 Format the four other subheadings in the same way.

8 Double space the paragraph headed *Viking Centre* only.

9 Save these changes and close the file.

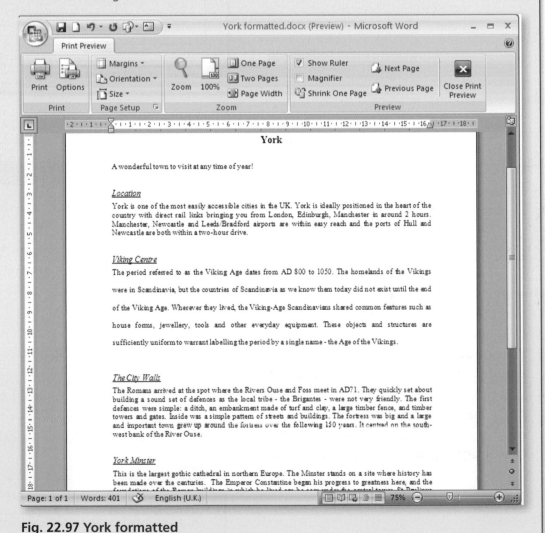

Fig. 22.97 York formatted

Margins

The normal Word 2007 document page margins are set at 2.54 cm top, bottom, left and right. To change any of these measurements, you can either select a pre-set option or enter exact measurements into the margin boxes in the **Page Setup** dialog box.

change margins

1 On the **Page Layout** tab, click on the drop-down arrow below the **Margins** button.

2 Select one of the styles offered.

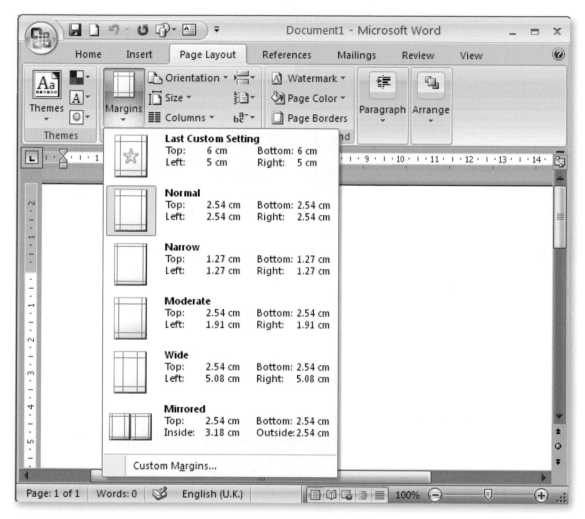

Fig. 22.98 Margins

3 To open the **Page Setup** dialog box, click on the **Custom Margins** link or the small arrow in the **Page Setup** group on the ribbon.

4 On the **Margins** tab you can use the up or down arrows or enter an exact measure in any of the margin boxes.

5 Note that 1/100th cm measures need to be entered manually into the margin boxes as you cannot set these using the up or down arrows.

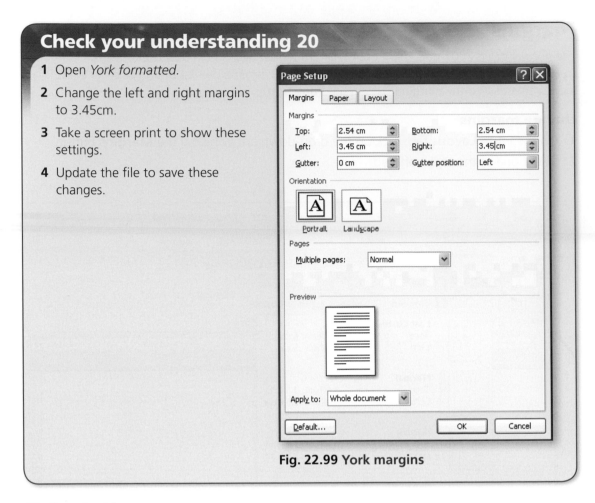

Check your understanding 20

1 Open *York formatted*.

2 Change the left and right margins to 3.45cm.

3 Take a screen print to show these settings.

4 Update the file to save these changes.

Fig. 22.99 York margins

Orientation

The default setting for Word 2007 is **Portrait** orientation where the shorter edges of the page are at the top and bottom. Some documents need to be **Landscape** to display the data more effectively with the longer sides top and bottom.

change page orientation

1 Click on the **Orientation** button on the **Page Layout** tab and select the alternative.

Or

2 Open the **Page Setup** dialog box and click on the correct button on the **Margin** tab.

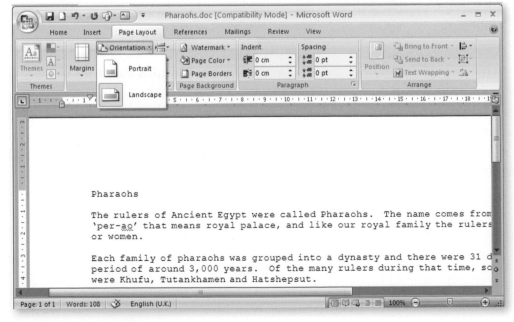

Fig. 22.100 Orientation

Page and paragraph breaks

If you want to break up a long paragraph into shorter ones, you can add a 'soft' paragraph break by clicking in front of the word that will start the new paragraph and pressing **Enter** twice.

By pressing **Enter** a number of times you could move any text that is to the right of the cursor onto a new page. But if anything is added or removed earlier in the document, the position of this text will change. If you always want a section of a document to start on a new page, you need to set a 'hard' page break. This will not be compromised by changes elsewhere.

In Normal view, you cannot see the page break after it has been set. To see it, click on the **Show/Hide** button on the **Home** tab. This displays all keystrokes and special characters such as spaces between words, tabs and paragraphs. A page break shows as a labelled, dotted line.

Show/Hide button

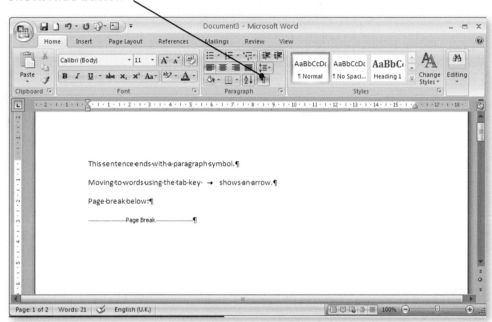

Fig. 22.101 Show hide

set a page break

1 Click in front of the text you want to start on a new page or click after the last paragraph.

2 Hold **Ctrl** and press **Enter**.

Or

3 Click on the **Insert** tab.

4 Click on **Page Break**.

5 Make sure the text starts on the first line of the page and at the left margin unless other alignments have been set.

6 To separate the current text from the new page, you could insert a **Blank Page** at this point.

Insert blank page   **Page break**

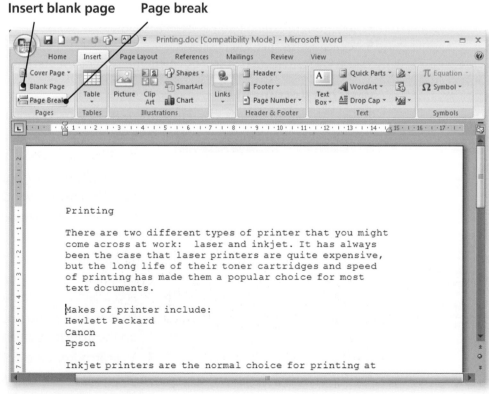

Fig. 22.102 Page break

remove a page break

1 Click in front of the first word on the new page.

2 Press the **Backspace** key one or more times.

Or

3 Click on the **Show/Hide** button, click on the dotted **Page Break** line and press the **Delete** key.

4 You may need to readjust spacing when the text moves back a page.

Check your understanding 21

1 Open the file *Printing* on the CD-ROM accompanying this book.

2 Create a new paragraph beginning **Many manufacturers**.

3 Insert a page break after the list of printers so that the text **Inkjet printers are the normal choice** starts on a new page.

4 Save as *Printing changes* and close the file.

Editing the text

If you are asked to revise or finalise a document, there are many facilities in Word 2007 that make the task quicker and easier. These include finding and replacing entries, inserting special characters and symbols and adding header or footer entries that will update automatically.

Using search and replace

As well as carrying out a simple search for a specific entry, or replacing an entry with another throughout a long document, the search and replace tools in Word 2007 allow you to carry out quite sophisticated searches or make changes to a document.

You can:

- find part or all of a word or phrase
- match case – to find entries that are case sensitive
- look for formatting that has been applied
- locate words that sound similar
- use wildcards – to find entries where the symbol * represents unknown characters.

search a document

1 On the **Home** tab, click on **Find**. The button shows binoculars.

Direction

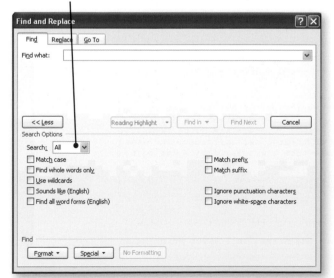

Fig. 22.103 Find

2 Enter the word or phrase you want to search for in the **Find what:** box.

3 Click on **More>>** (toggles with **<<Less**) to open the **Search Options** window if it is not visible.

4 Here click in the checkboxes as necessary to restrict the search.

5 To look for words with a particular emphasis or paragraph formatting, click on the **Format** button and select from the menus available.

6 To look for keystrokes, page breaks and so on, click on the **Special** button.

7 To search from the point you have reached in the document, select **Down** or **Up** in the **Search** box.

8 Click on **Find Next** to locate the first matching entry and repeat to work through the document.

replace entries

1 Click on **Replace** on the **Home** tab (the button shows *ab* ↪ *ac*) or click on the **Replace** tab in the **Find and Replace** window.

2 Enter the word or phrase in the **Find what:** box exactly as it appears in your document.

3 Enter the replacement entry in the **Replace with:** box.

4 To replace all entries, click on **Replace All**.

5 To locate and check an entry before replacing, click on **Find Next**.

6 To move on without replacing, click on **Find Next** again.

7 To replace a highlighted entry, click on **Replace**.

Check your understanding 22

1 Open the file *Italian* on the CD-ROM accompanying this book.

2 Insert a page break in front of the subheading *Rice*.

3 You need to replace entries of the word *rice* with *pasta*. Use the **Find and Replace** tool, but make sure you do not replace the subheading *Rice*.

4 Now carry out a search for an Italian sauce whose name you cannot remember. All you know is that it starts with *car* and ends with *a*.

5 Remove the page break you created earlier.

6 In the first paragraph, create a new paragraph beginning *This region...*

7 Save and close the file.

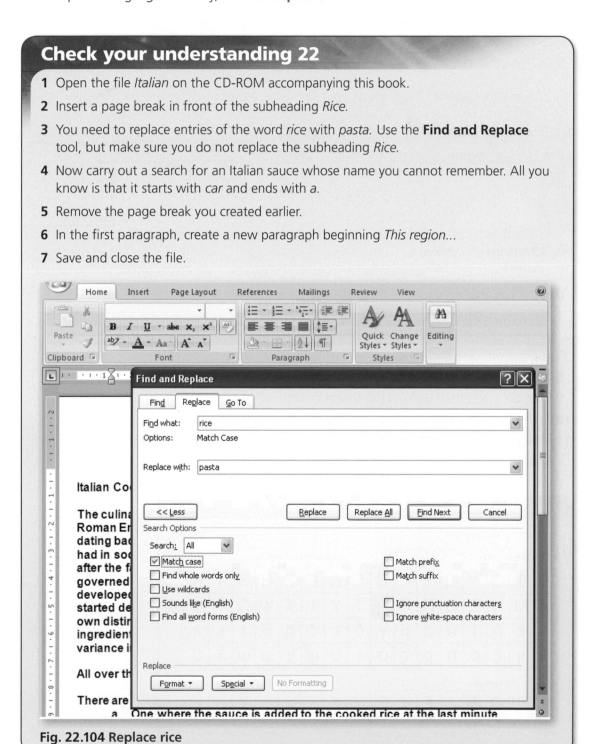

Fig. 22.104 Replace rice

Special entries

Some documents will require more than the normal letters, numbers or punctuation symbols visible on the keyboard. These include special symbols and characters such as Copyright © or Trademark ™. You may also need to change the currency symbol, for example to Euros €. All these symbols can be added from the **Insert** option.

insert symbols

1 Click on the page where you want the symbol to appear.

2 Click on the **Insert** tab.

3 On the far right of the ribbon, click on the **Symbol** button.

4 If appropriate, click on a recently used or common symbol showing here to add it to your text.

5 To find other symbols, click on **More Symbols**.

Fig. 22.105 Insert symbol

6 In the box that opens, recently used symbols will be readily available.

7 You may need to click in the **Font** and/or **Subset** box to choose from a different range of symbols but take care that, when inserted in the document, symbols are in the appropriate case and style.

8 Scroll down through the examples until you find the symbol you want.

9 Click on it and then click on the **Insert** button.

10 To insert Special Characters, find them by clicking on the tab.

11 Click on **Close** to leave the box and return to your document, or continue selecting and inserting more symbols.

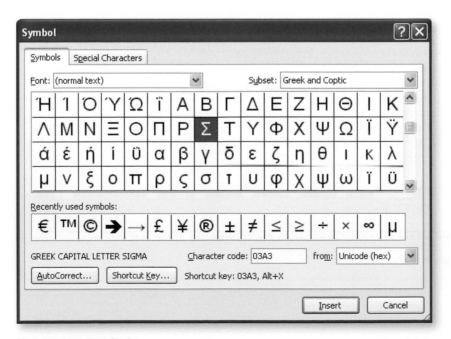

Fig. 22.106 Symbols

Note that although there are symbols available (for example, normal text – Latin–1 Supplement), to add European letters with accents such as acute, grave or umlaut, you can also use the keyboard.

type accented letters

1 Hold down **Ctrl**.
2 Press the key for any lower punctuation symbols such as **'** or **`**
3 Hold **Shift** and then press for upper punctuation symbols such as **^** or **:**
4 Type the letter:
   • Use **Ctrl 'e** for é
   • Use **Ctrl Shift ^ a** for â
   • Use **Ctrl `** (next to 1 on the keyboard) **e** for è
   • Use **Ctrl Shift :** (colon) **a** for ä

Typing equations and temperatures accurately

Instead of using symbols to type entries such as 35°C or C_6O_6 in a document, you format the characters so that they are smaller and sit above or below the normal typing line.

Characters above are **superscript** and characters below are **subscript**.

apply superscript or subscript

1 Type the letter or number as normal (for degrees of temperature, for example, use a lower case o).
2 Your entry might look like: **H2O** or **25oC**.
3 Select the relevant character with the mouse.
4 Click on the appropriate button in the **Font** group on the **Home** tab.
5 To continue typing normally if you have formatted the last character in this way, you will first need to click the button off.

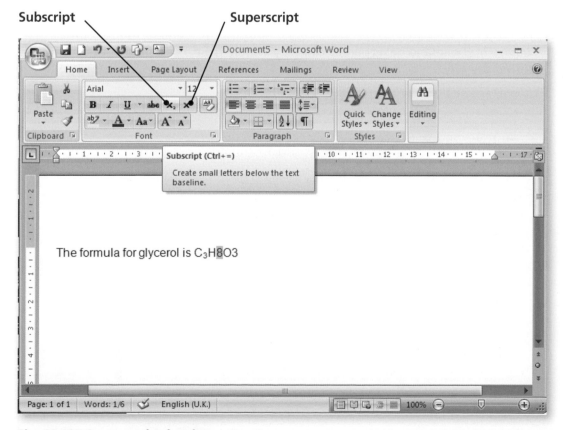

Fig. 22.107 Super and subscript

Click to prevent replacement

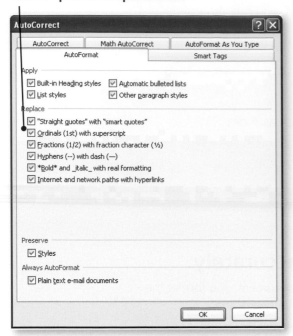

Fig. 22.108 Autocorrect options

Automatic superscript

You are likely to find that, as you type the ordinals 1st, 2nd and so on, superscript is applied automatically and entries become 1st, 2nd etc. This means that **AutoCorrect Options** have been applied. You can change the rules controlling such replacements by taking off ticks in the relevant checkboxes.

change AutoCorrect Options

1 Click on the **Office** button.

2 Select **Word Options**.

3 Click on **Proofing**.

4 Click on the **AutoCorrect Options** button.

5 On the **AutoFormat** tab, take off ticks in any boxes where you do not want the changes made automatically.

6 On the **AutoCorrect** tab you will find more rules and a list of replacements that you can delete or add to.

7 Click on **OK** to confirm any new settings.

Check your understanding 23

1 Open the file *Pharaohs* on the CD-ROM accompanying this book.

2 After the words *triangular tomb* in the third paragraph, insert a symbol to represent a pyramid or triangle (for example, from **Windings 3 Font** or **normal text – subset Box Drawing**).

3 At the end of the document, leave a clear line space and enter the following text, making sure the accents are included: **One famous French Egyptologist was Georges Aaron Bénédite (1857–1926) who investigated the Valley of the Kings.**

4 Start a new paragraph and type the following: **During the summer months the climate in the inland desert areas can vary widely from 7°C at night, to 43°C during the day.**

5 Save as *Pyramid*.

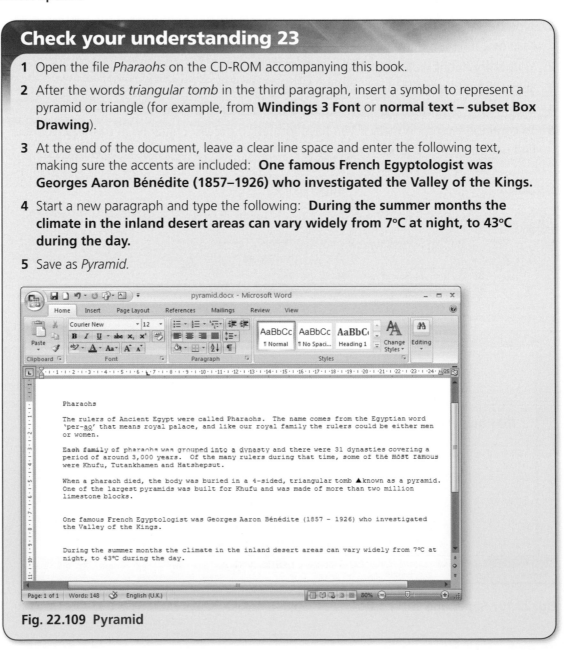

Fig. 22.109 Pyramid

Adding page numbers, date and time

It is very easy to add page numbers or the date or time to your documents. They are all available from the **Insert** tab.

number pages

1 Click on the **Insert** tab.
2 Click on **Page Number**.
3 In the gallery of styles you can choose a position and type of number.
4 Select from the **Page X of Y** section to include the number of actual pages in the document – useful when you want to indicate a document's length.

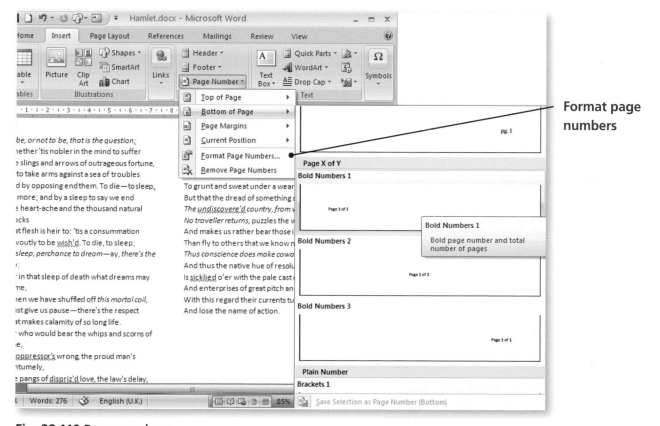

Format page numbers

Fig. 22.110 Page numbers

5 Click on the **Format** option if you want to change:
   • which page number the document will start at
   • the style of numbers.

add the date or time

1 Click on the **Insert** tab and select **Date & Time**.
2 When the window opens, select your preferred example.
3 Make sure the Language is set to English UK to put the day before the month.
4 Click in the update checkbox if you are delaying printing but want the current date on any document. (But note that, if you do this, the document will not display the date it was created when you refer back.)

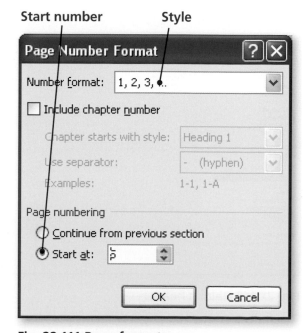

Start number   Style

Fig. 22.111 Page format

Headers and footers

Long documents often need a range of different information including the title or author added at the top or bottom of each page. These entries are known as **headers** (top) or **footers** (bottom). You can also add more specific information such as the file name or folder pathway. The value of using headers and footers is that they do not interfere with the main layout of the document pages. Also, entries such as date, time or page numbers inserted from the menu can be set to update automatically.

To add several items, move across the header or footer box by pressing the **Tab** key or double clicking the mouse, and move onto a new line by pressing **Enter**.

add headers and footers

1 Click on **Header** and select a style such as **Blank**.

2 When dotted lines show the position for the header box, start making entries.

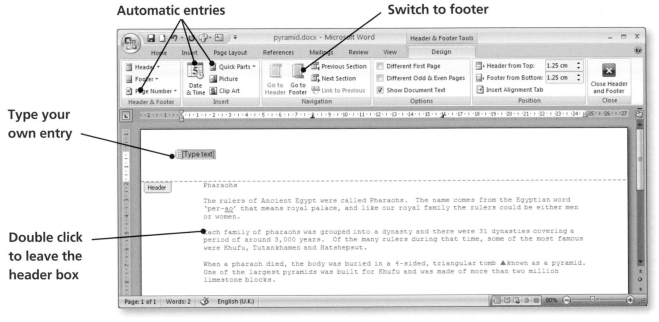

Fig. 22.112 Header toolbar

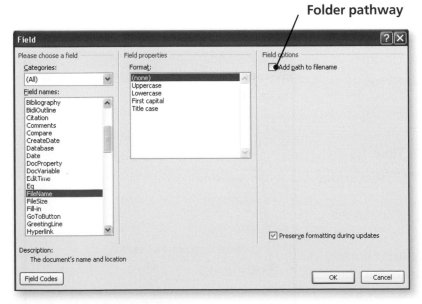

Fig. 22.113 Field in header

3 Click on buttons on the **Header and Footer Tools – Design** tab to add extra fields.

- **Page Number** offers styles to choose from.
- **Date & Time** will open the dialog box where you can choose how the date/time entry will be formatted. Click on the checkbox to update the entry automatically.
- **Quick Parts – Field** will open the list of fields. Scroll down and click on **FileName** to add this. Click on the checkbox if you also want to show the folder pathway and click on **OK** to return to the document.

4 To make entries in the footer, click on the **Go to** button.

5 To return to normal typing, double click on the greyed out document text or click on the **Close Header and Footer** button.

6 Reopen the header or footer box by double clicking on an entry.

Bullets and numbering

Lists can be enhanced and the contents are often displayed more clearly if **numbers** or **bullets** are added. The style of number or bullet can be changed by selecting a different option, and it is easy to remove them if they are unwanted.

If you apply bullets or numbering before typing, these will be added each time you press **Enter**.

When bullets or numbers are added, the text will be indented automatically and you can change the distance that both the bullets and the text are from the left margin.

add bullets or numbers

1 Select list items or follow these steps first.

2 Click on the **Bullets** or **Numbering** button. The default style will be applied.

3 To change this, click on the drop-down arrow next to the button and select an alternative.

4 Take off bullets or numbering by clicking on **None** or selecting the list and clicking off the button.

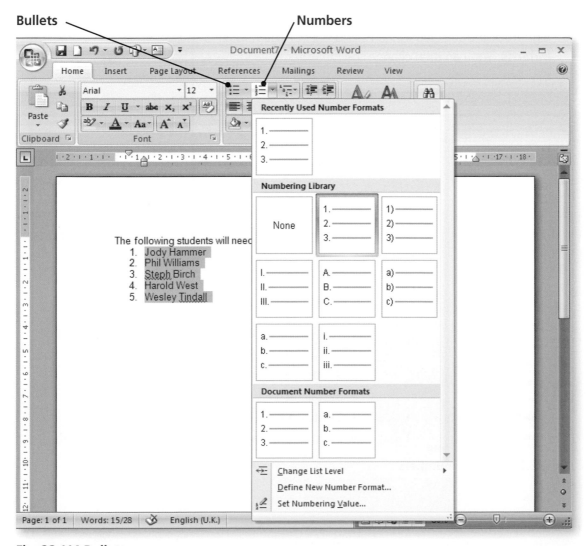

Fig. 22.114 Bullets

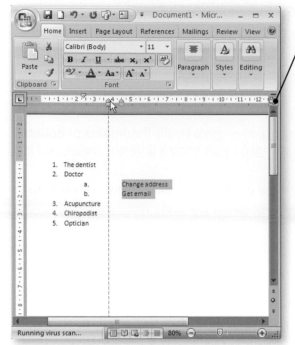

View ruler

Fig. 22.115 Number ruler position

5 If text or a bullet or number is in the wrong position on the page, select the list item and then drag the relevant marker across the ruler. (Click on the **View Rulers** button at the top of the vertical scroll bar if it is not displayed.)

Or

6 Right click the selected list items and select **Adjust List Indents**. You can then change the measurements in the various boxes.

Fig. 22.116 List indent box

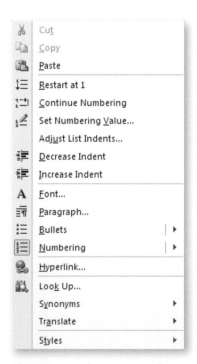

Fig. 22.117 Change numbers

7 To change the numbering, for example if a list is split across an image, right click on one of the numbers and select the appropriate menu option:
- Restart at 1
- Continue from an earlier number
- Change the start number (**Set Numbering Value**).

Check your understanding 24

1 Start a new document.

2 Type the following text exactly as set out below:

Countries I have visited

Over the past few years I have visited a number of different countries and these include:

Sweden
India
Mexico
Germany
Ireland

My favourite place was Stockholm.

3 Save as *Countries*.

4 Apply bullets to the list of countries.

5 Change to a different style of bullet.

6 Double space just the list.

7 Add a footer showing the page number and file name.

8 Close and then reopen the file.

9 Add the word *Geography* as a header.

10 Change to landscape orientation.

11 Save and close the file.

Fig. 22.118 Countries

Outline numbering

It is often necessary to include sub-list items at a lower level. Where more than one level of list item is included, this is known as outline numbering. The numbers will have a different format and indentation at each level and Word 2007 will automatically set a new level if you use the **Tab** key when creating a list.

apply outline numbering

1　For a list that is already typed, apply numbering to the entire list.

2　For any line you want at a sub-list level, click in the line and press the **Tab** key.

3　Repeat this to move down to the next level.

4　Hold **Shift** as you press the **Tab** key to move the line up a level.

5　To create a new list, start numbering as normal.

6　Press **Enter** to type in the next item.

7　For a sub-list item, press **Enter** and then press the **Tab** key.

8　Type the entry.

9　For a new entry at the same level, just press **Enter**.

10　To move up to a higher level for the next item, press **Enter** and then hold **Shift** as you press **Tab**.

11　You can select a different style of multilevel list if you open the gallery.

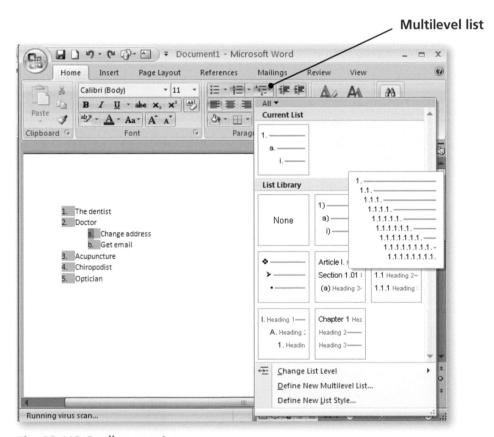

Fig. 22.119 Outline numbers

12　If the style you want is not visible, click on **Define New Multilevel List** to open the dialog box. Select any list item level and choose a different style of numbering.

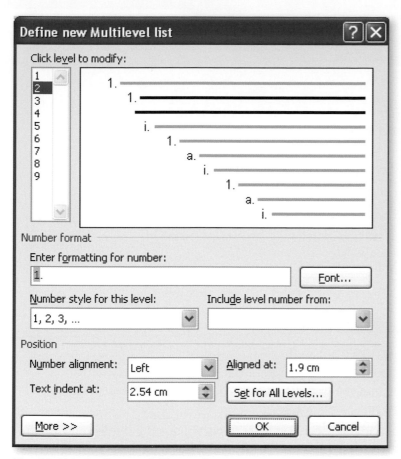

Fig. 22.120 Define multilist style

Check your understanding 25

1 Open *Holidays* on the CD-ROM accompanying this book.

2 Recreate the list as shown below using the style of numbers that appear automatically.

3 Save as *Lists* and close the file.

4 Reopen the file and change the second level of numbers to a different style.

5 Save and close.

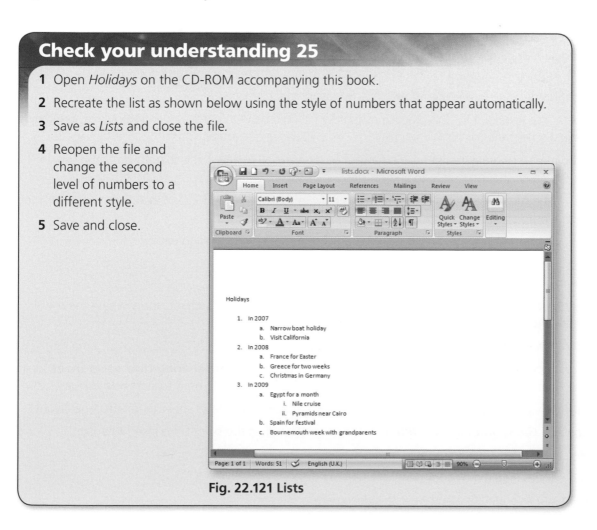

Fig. 22.121 Lists

Assignment

This practice assignment is made up of five tasks

- Task A – Design and create a newsletter template
- Task B – Create a new document from newsletter template
- Task C – Design, create and merge a letter
- Task D – Use word processing tools
- Task E – Saving and printing

You will need the following files:

- Call Of Duty for PS3.doc
- New Super Mario Bros for Wii.doc
- Newsletter text.doc
- Game-sharing club graphic.jpg
- Game sharing club logo.jpg
- Game sharing club photo.jpg

Scenario

You have realised that most of your friends, near and far, own and enjoy similar games consoles. During an enjoyable night out a few days ago, it was decided that you'd start a game sharing club so that expensive games could be passed on and swapped around in the group.

It was decided that you would set up a news sheet and accompanying letter to distribute around the group listing available games cartridges as well as news and views on new products in the marketplace.

Please read the text carefully and complete the tasks in the order given.

Task A – Design and create a news sheet template

1   Create a new document named **practice assignment evidence** and save it to a new folder named **practice assignment**.

   You may add to this document as you complete practice assignment tasks.

2   The text and data files you need for this practice assignment should be in a folder named **WP**. Make sure they are present, take a screen shot of them and paste into your evidence document.

3   Use pencil and paper to create a design for the news sheet to be used by the game sharing club. The design should include the game sharing club logo and articles shown in Appendix A.

   The paper size will be A4. The orientation can be your choice of portrait or landscape. Use a ruler and pencil to show the sizes of your margins.

   Include two columns for the main body text with the articles. Annotate your design with:

   a) Alignments
   b) Names of fonts, their sizes, and any enhancements or styles.

   There will need to be sections between the columns and other parts of the document; show where these will be using the ruler and pencil.

   Include a footer with the document filename and print date.

4   Create a new word processing document. Save the document, naming it **Game sharing club news sheet**, as a template to the Trusted templates (Word 2007) or equivalent folder, so it can be used to start new documents.

   Use the pencil and paper draft plan you created in Task A as a guide to this news sheet template page layout.

5   Insert the club logo, **Game sharing club.JPG**, into the template, adjust the size of this inserted object.

6   Create a footer, with suitable fields for the document name and print date.

7   Insert section breaks as required by your design.

8   Save the page layout as a template and close this document.

Task B – Create a new document from news sheet template

1   Create a new document based on the **Game sharing club news sheet** template you created in Task A3.

Insert the text from and graphics from:

- Call Of Duty for PS3.doc
- New Super Mario Bros for Wii.doc
- Game-sharing club graphic.jpg
- Game-sharing club photo.jpg

Format the document so it matches the design and looks professional.

2   Save the document as *GSC news sheet RJM*, where RJM represents your initials.

3   Print the document.

Task C – Design, create and merge a letter

1   Read this practice assignment to see the contents of your news letter, then type into your evidence document a suitable sub-heading with a brief plan for the production of the mail shot.

The plan should include:

a) The mail-merge data source
b) Templates needed for the news letter
c) Any graphics needed for the news letter
d) Consumables needed for printing

2   Use pencil and paper to create a design for the news letter to be used by the game sharing club.

The paper size will be A4. The orientation will be portrait. Use a ruler with pencil to show the sizes of your margins.

Annotate your design with:

a) Alignments
b) Names of fonts, their sizes, and any enhancements or styles
c) Merge fields and what data they will show

There will be a footer with the filename and print date.

Plan which fields will be suitable for the structure of your data file.

3   Create a new document using a word processor. Save the document, **Game sharing club news letter**, as a template to the Trusted templates (Word 2007) or equivalent folder, so it can be used to start new documents.

4   Use the pencil and paper draft plan you created in Task C1 as a guide to this news letter template page layout.

The text can be found in supplied document, *Newsletter text.doc*.

Format the text and table to make it easily readable and look professional.

5   Create a footer, with suitable fields for the document name and print date.

6   Use the spell checker changing mis-spelt text. Add the names of any club members found as mis-spellings to the spell checker dictionary.

7   Proof read the document, carefully check that the text is accurate and correct.

8   John House is no longer the club leader and has been replaced with Jane Howse. Use search and replace to change all occurrences of John House to Jane Howse.

9   Insert the merge fields into this document.

10 Save the page layout as a template and close this document.

Task D – Use word processing tools

1   Create a new document based upon the **Game sharing club newsletter** template you created in Task C.

2   Create a data file and accurately input this required data:

Bill Bayleigh	12 High Street	Bristol	BS1 3RD
Rachel Haynes	3 Otter Road	Keynsham	BS37 2FG
Steve Starr	1 Compass Way	Bristol	BS8 1QS
Mary Wand	34 Broadway Road	Bristol	BS5 5TT

3   Merge these documents and preview the results to your screen.

4   Selectively merge the data so only the addresses in Bristol are used.

Task E – Saving and printing

1   Save your mail-merged document as **GSC newsletter RJM** (using your initials instead of RJM) into:

a) My documents

b) A removable/portable media such as a pen drive.

2   Add paper to the printer as necessary. Preview your documents, make any changes you think useful and print.

Check your printed outputs for accuracy and layout.

Close finished documents and the word processing application.

Appendix A

Artwork available for the games sharing club news sheet and letter:

Game sharing club

Game-sharing club logo

Game-sharing club graphic

Mario simultaneous multiplayer for the Wii is here!

Up to four can play on the same level at the same time at any point in the game for competitive and cooperative multiplayer fun.

Players can pick each other up to save them or toss them into danger.

Mario, Luigi and two Toads are all playable characters, while many others from the Mushroom Kingdom make appearances. Players can ride Yoshi characters and use tongues to swallow enemies – or their fellow players.

The motion abilities of the Wii Remote controller can tilt a seesaw to help a character reach a higher platform or tilt it incorrectly to delay other players.

The new propeller suit can shoot players high into the sky with just a shake of the Wii Remote!!

During a simultaneous multiplayer game, players are ranked on their score at the end of each stage, with coins collected and the number of enemies defeated.

http://hmv.com/hmvweb/displayProductDetails.do?ctx=1008;6;-1;-1;-1&sku=991326&WT.ac=Chart+Games-PBODY-Chart_Games-991326

Game-sharing club photo

Another great first person shooter series returns to the modern day.

With Call of Duty 4: Modern Warfare, you play Sgt Gary "Roach" Sanderson. Your commander is "Soap" from the first game and your mission is to stop Russian ultra-nationalists.

Your squad builds on experience, developing from inexperienced rookies to veteran warriors.

The game features a more varied series of locales than ever before, including a snowbound enemy camp. The adventure begins with a daredevil mountain climb across the ice before a stealthy infiltration goes awry and you have to commandeer a snowmobile to run for your life.

Stunning cinematic action includes new underwater missions, brand new gadgets and more vehicles. You may travel to the deserts of Afghanistan, the slums of Rio de Janeiro and even the wilderness of Russia...

http://hmv.com/hmvweb/displayProductDetails.do?ctx=1008;6;-1;-1;-1&sku=915443&WT.ac=Chart+Games-PBODY-Chart_Games-915443

IT communication fundamentals/Using the Internet

This unit will help you develop a more in depth understanding of the Internet and browser technology. It also covers the computer operating system, IT security issues and provides an introduction to email and different types of electronic network.

In particular, you will be able to:

⊕ identify system requirements

⊕ describe and use Internet services

⊕ use email

⊕ use internet conferencing

⊕ identify Internet security issues.

Internet connections

As you learned at Level 1, there are five things you need in order to access the Internet:

- computer
- telephone or other connection
- **modem** or **router**. A modem is hardware that converts the digital computer signals to analogue so they can be sent via the telephone system. Most modern computers come ready fitted with an internal modem. A router is needed if you have networked computers, as information can then be directed or 'routed' to specific machines
- **Internet Service Provider (ISP)**. These are organisations you will have an account with and which provide access to the Internet
- **browser**. This is the software required to view web pages. Common browsers include Internet Explorer, Firefox and Opera.

Connections to the Internet are changing all the time, but you might use one or more of these:

- telephone system
- mobile phone
- cable
- satellite
- WiFi
- mobile broadband 'dongle' plugged into the computer's USB
- games consul

Originally, computers made use of dial-up (analogue) services across the telephone system, but now the speed and versatility of broadband means it is the main choice for most Internet users.

Connection speeds

Internet connection speeds are measured in the number of seconds it takes to receive thousands of bits of data – **Kilobits per second (Kbps)** or **Megabits per second (Mbps)**.

- Dial-up – Typically these range from 2400 Bps to 56 Kbps.

- ISDN – This stands for **Integrated services digital network** which is an international communications standard for sending voice, video, and data over digital telephone lines or normal telephone wires. Typical ISDN speeds range from 64 Kbps to 128 Kbps.

- DSL – This uses existing two-wire copper telephone line and allows normal phone use. DSL is always on and can be one of two types: **ADSL** (asymmetric digital subscriber line) is used in the UK and USA, and **SDSL** (symmetric digital subscriber line) is used more in Europe. ADSL supports data rates of from 1.5 to 9 Mbps when receiving data (known as the **downstream rate**) and from 16 to 640 Kbps when sending data (known as the **upstream rate**). It requires a special ADSL modem.

- Cable – with a cable modem you can have a broadband Internet connection that is designed to operate over cable TV lines as it uses TV channel space for data transmission. It has a greater bandwidth than telephone lines and so can be very fast, ranging from 512 Kbps to 20 Mbps.

- Fibre optic broadband ISDN transfers data over fibre optic telephone lines, not normal telephone wires, and so these have to be laid in. It is not currently used to a great extent, although in 2010 several UK cities were beginning to experiment with it.

- Wireless broadband uses radio frequency bands. Like DSL it is always on and can be accessed from anywhere within the network coverage area. It may seem slow if many people are accessing the same channel and there is interference, and it requires the wireless card to be configured correctly.

- Leased lines – These are a business option where a dedicated phone connection supports huge data rates. Separate channels each supporting 64 Kbits per second can be configured to carry voice or data traffic and may be bonded together to increase bandwidth.

- Satellite allows you to access the Internet via a satellite above the earth's surface. As the signals must travel an enormous distance, this has a slightly lower connection speed than copper or fibre optic cables and so speeds are from around 492 to 512 Kbps.

Did you know?

There is no perfect way to connect, as each method has benefits and drawbacks relating to connection speed, the ISP's customer service, stability of connection, and accessibility. The connection you want might not be available in your area.

Selecting an ISP

When setting up your own Internet access, you will need to choose an appropriate service provider. As there are so many companies offering this service, the choice is very wide. The best for you personally will depend on a number of factors:

- How much will you be using the Internet? Dial-up is cheaper than broadband but very limited when it comes to accessing video, music or large images.

- Do you want a combined package with TV and telephone? Some companies offer good value when you take several services.

- How much downloading will you carry out? There may be a limit on what you can download, with heavy users paying more.

- Do you want help if you have problems? Some help desks are very expensive to phone up.

- What speed do you require? Different types of connection have very different speeds.

- Do you have a telephone or cable connection? If not, mobile or WiFi may be the only options.

Remember

You can often use your ISP's online help facilities to solve connection problems. However, if you have no connectivity, you won't be able to access online help.

Check your understanding 1

1 Visit www.broadband-finder.co.uk or another broadband comparison site.

2 Compare the different types of service being offered.

3 Find out the price you would need to pay for a year's subscription to a broadband service that would suit you personally.

Customising a browser

Each time you open your browser to view pages on the Web, it opens at the same web page, known as your home page. This is usually set as a page on the website of your ISP or you may have changed this by mistake when clicking on a link within a page you visited.

If you prefer to start an Internet session at the original page, or with a search engine or a website that you regularly need to display, you can set a different home page using the browser's **Tools** menu.

You can also change other settings if you want to customise exactly how your browser operates. For example, you can add, remove or reposition toolbars, prevent images downloading (which slow down web browsing), set access restrictions and deal with temporary files.

Single tab

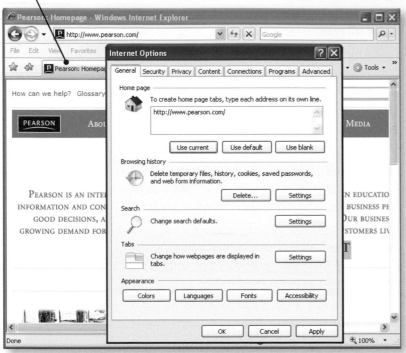

Fig. 25.1 Home page

Drag or move here

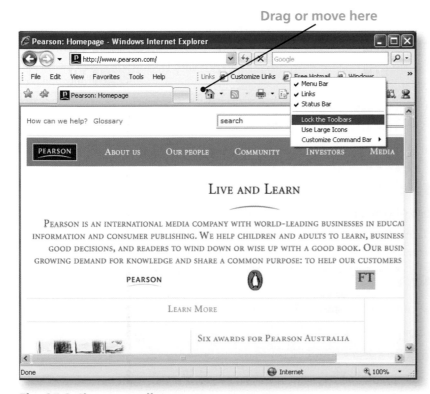

Fig. 25.2 Change toolbars

set a new home page

1   Navigate to the page you want as your home page.

2   Go to **Tools – Internet options**.

3   In the Home page section, click on **Use current** to replace the entry with the URL of the page you have opened.

4   If you have set more than one tab to open in your browser, you can add new addresses in the box for the other tabs.

5   Click on **OK** and this page will open next time you start up the browser or click on the tab.

change toolbars

1   Right click on any toolbar for the list of toolbars you could display.

2   Click on any name to add it to the screen.

3   To move toolbars above or below one another, make sure they are unlocked by taking off the tick on the **Locked** option. The vertical edges will appear.

4   Now move the mouse to the left-hand edge of any toolbar, hold down the mouse and drag it to a new position when the pointer becomes a four-way arrow.

5   If the tools are not clearly visible because they are too far to the right, position the pointer over a dotted vertical line marking the left-hand edge of the toolbar and drag this to the left when the pointer shows a two-way arrow.

6   Click on any toolbar on the list that has a tick to remove it.

7   If you prefer, lock the toolbars into their new positions.

enable/disable images

1 Go to **Tools – Internet Options**.
2 Click on the **Advanced** tab.
3 Scroll down through the list of settings and add or remove ticks in checkboxes related to **Multimedia**. You can:
- stop pictures being displayed
- show a placeholder indicating the position for a picture
- show pictures
- stop animated pictures running.

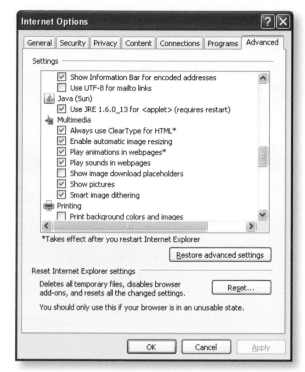

Fig. 25.3 Enable images

Web page appearance

There are a number of reasons why the same web page viewed on your computer may look different from that on another machine.

- You may be using a different browser. Although similar, some do offer quite a different display. This may be important if you ever publish your own web pages and want all visitors to have the same experience.

- Your computer screen resolution. This is a measure of the pixels or dots of colour that make up the picture. The higher resolution will improve the sharpness of the image. A high resolution might be 1024 x 768 pixels, where icons and windows will be at their smallest. When decreased to 800 x 600, objects are blocky and much larger.

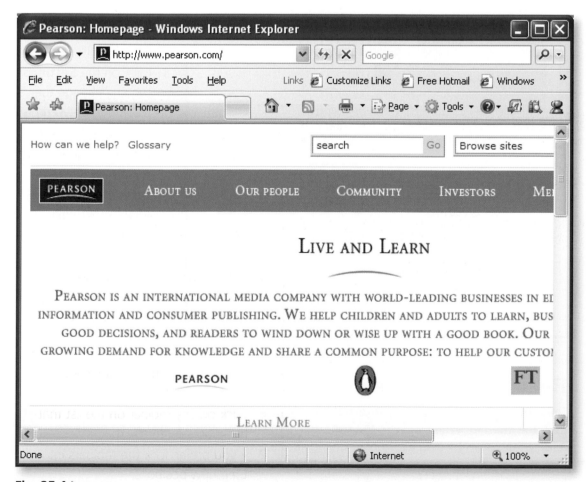

Fig. 25.4 Low res

Fig. 25.5 High res

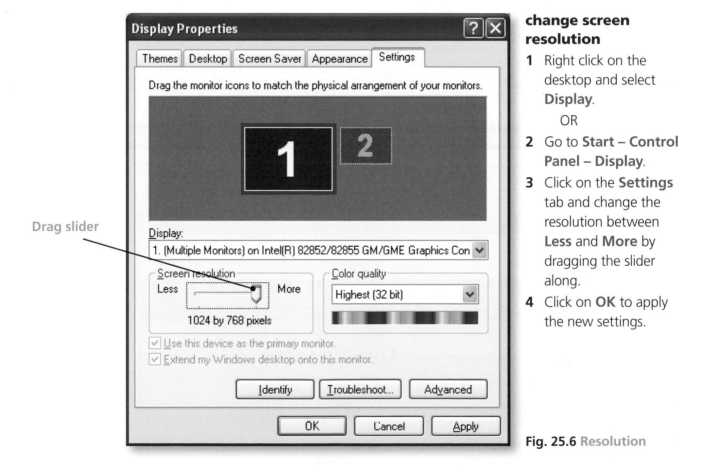

Drag slider

change screen resolution

1 Right click on the desktop and select **Display**.
   OR
2 Go to **Start – Control Panel – Display**.
3 Click on the **Settings** tab and change the resolution between **Less** and **More** by dragging the slider along.
4 Click on **OK** to apply the new settings.

Fig. 25.6 Resolution

Recognising the relevance of cache settings

The browser cache (**Temporary Internet Files** folder) stores pages locally that are downloaded from the Internet. When you visit the page again, the locally stored page can be displayed more quickly than downloading a new one. Changing the cache settings depends on your Internet activity. If you browse a great deal, increasing the size of the cache means it can store more temporary files, but this storage will leave less space for other files and so now and again the cache should be cleared.

change cache settings

1 Go to **Tools – Delete Browsing History**.
  Or
2 Go to **Tools – Internet Options**.
3 On the General tab, under Browsing History click on **Delete**.
4 Click on **Delete files** under **Temporary Internet Files**.
5 Click on **OK** to confirm the deletion.
6 If you click on **Browsing History – Settings** on the **General** tab you can increase or reduce the amount of space allocated to temporary files.

Configuring web page access restrictions

Many web pages have content that you may not want to view, or allow other users of your computer such as children to view. Internet Explorer has a program known as the **Content Adviser** that consists of ratings that can be set to limit what is viewed by their general content.

You can also add specific web page addresses to the list of approved sites so that these pages can still be accessed despite appearing to meet the requirements of rules that would normally prevent their display.

use the Content Adviser

1 Go to **Tools – Internet Options**.
2 Click on the **Content** tab.
3 Click on **Enable** under **Content Advisor**.
4 On the **Approved Sites** tab, add any websites that you always want to be accessed (or denied) whatever ratings are set.
5 On the **General** tab, click to set up a **Supervisor** password that will allow you to change the ratings for various types of website access and turn the **Content Advisor** on or off.

Fig. 25.7 Delete browsing

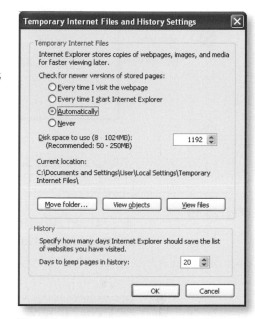

Fig. 25.8 Cache settings

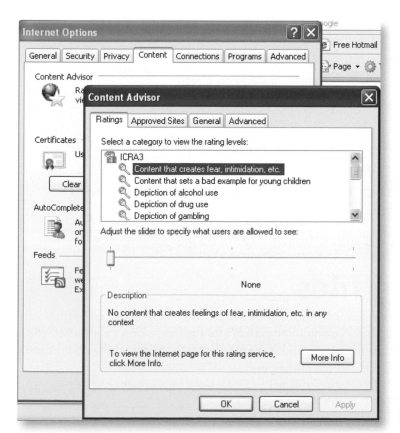

Fig. 25.9 Content access

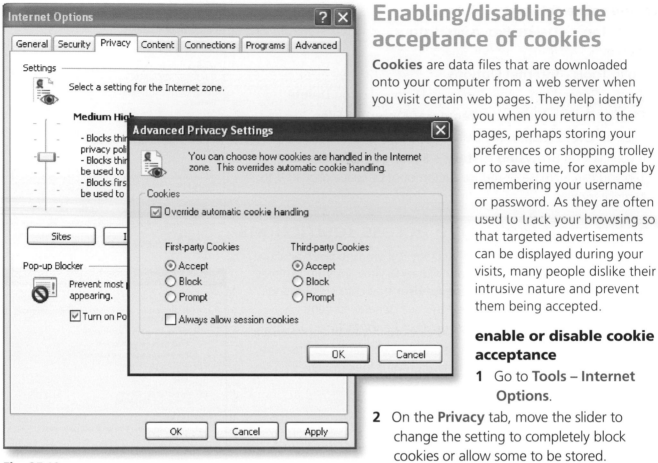

Fig. 25.10
Cookies

Enabling/disabling the acceptance of cookies

Cookies are data files that are downloaded onto your computer from a web server when you visit certain web pages. They help identify you when you return to the pages, perhaps storing your preferences or shopping trolley or to save time, for example by remembering your username or password. As they are often used to track your browsing so that targeted advertisements can be displayed during your visits, many people dislike their intrusive nature and prevent them being accepted.

enable or disable cookie acceptance

1 Go to **Tools – Internet Options**.

2 On the **Privacy** tab, move the slider to change the setting to completely block cookies or allow some to be stored.

3 Click on **Advanced** and click in the **override** checkbox if you want to allow certain types of cookie such as session cookies that are not stored but are always deleted after a visit.

Check your understanding 2

1 Change your home page.

2 Check that the new setting has taken effect.

3 Change back, if necessary, to the original home page.

4 Add a new toolbar and manually change its position on screen.

5 Explore the list of settings you can change on the **Advanced** tab related to browsing, printing and security.

6 Check whether cookies have been blocked on your computer.

7 View the same web page using two different browsers.

8 Change your screen resolution and see what effect this has on a web page being displayed.

Compressing files

One way to save space is to store copies of important files in a compressed or zipped format as an archive file. The archive can then be saved on removable storage media or sent attached to an email. When an archive is received or when you want to access the contents, the compressed files can be extracted and then treated in the normal way.

Windows XP or later machines contain the software to archive files directly. For earlier operating systems, you will need to use a program such as **WinZip** that is available free on the Internet.

zip files using Windows XP or later

1 On the desktop, select all the files you want to compress. Hold **Ctrl** as you select individual files if they are not adjacent.

2 Right click on any selected file and select **Send to – Compress (zipped) Folder**.

Fig.25.11 To zip files

3 A new yellow folder containing the compressed files will appear. The icon has a zip across it, and it will be labelled with the same name as one of the selected files. The file extension will be **.zip**.

4 Rename the archive to avoid confusion.

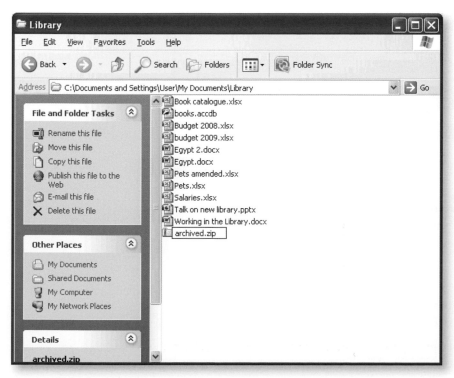

Fig. 25.12 Rename archive

5 If you double click on the archive, you will display copies of all the selected files.

6 You will also be able to see that the whole archive takes up less space than the individual files would normally.

Fig. 25.13 Opened archive

work with archived files

1 Open the **archive**.

2 Double click on an individual file to open it as normal.

3 To save all the files outside the archive, click on the *Extract all files* link in the **Folder Tasks** pane. This starts the **Extraction Wizard**.

4 Click on **Browse** to select a destination for the files. At this stage, you could create a new folder to contain the extracted files. Right click on it when it appears in the folder list to give it a suitable name.

Fig.25.14 Extracting

5 Click on **Next** and watch the progress of the files being extracted.

6 Click on **Finish** and choose whether to view the extracted files at this stage or not.

Check your understanding 3

1 Select any three files saved in **My Documents**.

2 Archive them and name the new archive *Compressed*.

3 Open the archive to check its contents.

4 Now extract the contents to a new folder named *My zipped files*.

5 Open this folder and take a screen print of the contents.

Fig. 25.15 Archive in my documents – Step 2

Fig. 25.16 My zipped files – Step 5

Navigating the Web

To find information, you have to open a relevant web page. This can be carried out in two different ways:

- Use the full **URL** by typing it into the **Address** box. You don't need to start with http://, and sometimes not even www, because if you have visited the page in the past its address will be stored in your browsing history. As you start typing the name of the company or organisation, the rest of the address is displayed automatically.

- Click on one of the links placed on an open page. These are known as **hyperlinks** and will open a related page. They are usually identified by coloured text (although images do not appear any different) and the pointer changes to a hand when over any hyperlink image or text.

Once you have stored the URL of a page you may want to revisit in your folder of favourite pages – known as **bookmarking** – you can open that page by clicking its name on screen.

Searching the Web

For many people, the Web is a valuable tool for buying goods or services. This type of electronic shopping has been given the general name **e-commerce** and most large organisations such as supermarkets, banks, train companies, DIY outlets, airlines and so on who want to advertise or offer goods or services to customers will have a presence on the Web. The official website is clearly the place to visit if you want to buy from a named company, and so you simply need to type in the URL to view their pages and make your purchases.

If you want to compare prices across a number of outlets, take an online training course, get the latest news or weather forecast, watch videos or research a range of topics, it won't be clear which website contains the material you are looking for. In such cases, you need to visit a site known as a **search engine** which has been created specifically to help you search the Web for relevant information.

The words you type – known as **keywords** – will be used to locate web pages that contain the same words. Although many sites listed will not be relevant, a good search engine should identify at least a few websites worth following up.

Some browsers already offer a limited search facility, but it is usual to visit a search engine site as you can then customise your search.

The following addresses are a few of the many search engines or **metasearch engines** (they search the search engine sites) that you can try:

www.google.co.uk	www.yahoo.co.uk	
www.ask.com	www.metacrawler.com	www.bing.com

Effective search techniques

With so much information available, it is a good idea to try to reduce the length of a search list so that you only have a short but relevant group of websites to visit. Here are a few ways to do this:

- Click on a link on the page or add *UK* as a keyword to limit the search to UK websites only, if this is important.

- Click on an Image, Maps or other special link if you want to search for these rather than carry out a general website search.

- Type several words between quotation marks to limit the results to pages that contain those words next to each other.

- Enter a whole sentence if you are not sure which keywords to select – search engines ignore *and, the* etc.

- You can add a plus **+** sign in front of words that must be included, or a minus **-** sign to omit pages containing specific words. You can also try using the word **NOT** to exclude certain pages.

- Use **AND** or **OR** so that, for example, you find sites that contain both keywords you are searching for or only alternatives.

The Internet is full of information, but making the most of that information can be difficult. As a user, you will need to learn to identity how well information meets requirements.

First, is the information up to date? If you are searching for population data, and the page you have found gives you figures from the 1970s, that information will not be relevant today.

Second, where has the information come from? Is it unbiased facts, or did the creator of the piece have an agenda to push? A vegetarian society and a cattle farmer will have very different opinions on the health benefits of beef.

There are many factors to consider: is the level of detail sufficient for your requirements? Is the data relevant to your search query? Does the information have an obvious bias? Was it provided by someone with an agenda?

The best way of ensuring you are making the most of the information you have in front of you is to gather data from a range of sources and cross-check it for accuracy. This is also known as synthesising information.

Check your understanding 4

1 Visit the search engine site www.ask.com.

2 Use keywords to find as many recipes as you can using chicken liver and note how many sites have been found.

3 Now limit the search to chicken liver paté and again check the number of sites.

4 Visit two different sites and check that they contain relevant information.

5 Note down the full URLs of each site.

Bookmarking pages

No one can retain in their heads the full URL of all pages visited, so browsers offer users a folder where they can store a link to any page they come across when browsing that they might want to revisit. This is known as **bookmarking** the page. In Internet Explorer the folder is known as the **Favourites** folder.

bookmark a page

1 With the page open in the **browser** window, hold **Ctrl** and press **D**.
 Or

2 Click on the **Add to Favourites** button showing a yellow star and green cross.
 Or

3 Open the **Favourites** menu and click on **Add to Favourites**.

4 Retype the page name if it is too long, and click on the **Add** button. This will store the URL at the bottom of the list of favourite websites.

5 If you prefer, you can first select an appropriate subfolder in which to store the URL from the **Create in:** box. You could also create a new folder at this stage by clicking **New Folder** and entering a suitable name.

6 Click on **Add** to place the URL inside the folder and close the box.

Add favourites button

Create a new folder to store the URL

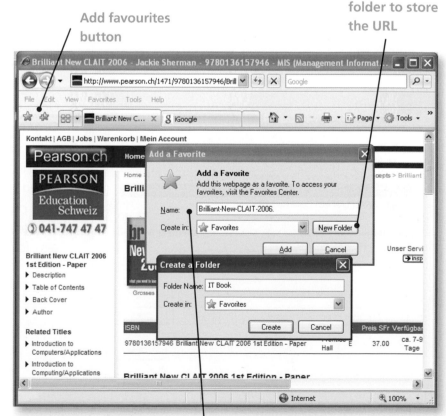

Fig. 25.17 Add favourite

Amend page name

visit a bookmarked page

1 Click on the **Favourites Centre** button.
2 Click on the folder containing the bookmark and then click on the **URL**.
3 The page will now open.

Favourites centre

Click to open the page

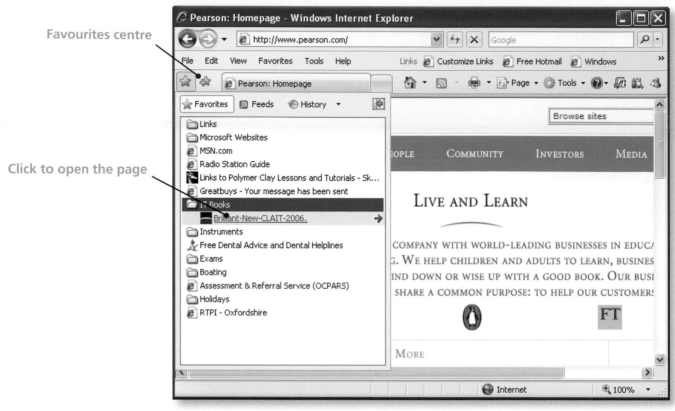

Fig. 25.18 Open bookmark

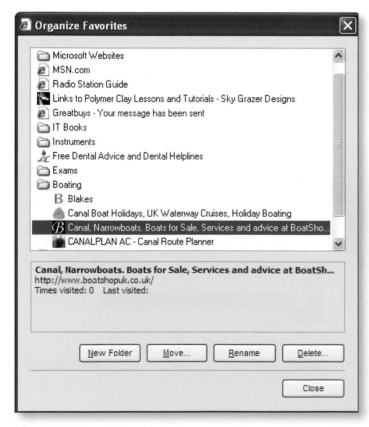

Fig. 25.19 Organise

manage favourites

1 On screen, open the list of favourite folders.
2 Drag a page up or down the list using the mouse.
3 Right click to delete or rename a page.
  Or
4 Open the **Favourites** menu and select **Organise Favourites**.
5 In the window that opens, select a page. You may first have to open the folder in which it is stored.
6 Now click on the appropriate button: move it to a different folder, rename or delete it or create a new folder.

Check your understanding 5

1 Use any search engine to find details of what is on this week in your local cinema.

2 Bookmark the page in a new folder named *Cinema*.

3 Now bookmark a page displaying information about Odeon cinemas in London's West End and store it in the same folder.

4 Finally carry out a search to display information about who played Proximo in the film *Gladiator.*

5 Bookmark the page and store it in the *Cinema* folder.

6 Use your bookmarks to reopen the local cinema page.

7 Finally, delete the link to *Gladiator.*

Sharing information

One way to find out about things over the Internet is to share knowledge with other people through email or publishing messages online. You might use:

- **Social networking sites** such as Facebook or Bebo – these work by allowing you to publish a personal profile and other users can then send you messages that will appear on your page.

- **Discussion forums (or bulletin boards)** – here members normally offer advice or complain about things. Messages are grouped into various topics and you send in written comments or simply read what others have written.

- **Chatrooms** – these allow you to communicate in real time. Other people need to be online at the same time and you have written conversations by typing into a box.

- **Newsgroups** – these consist of groups of people with the same interest who communicate via email. You set up your email system with newsgroup folders that will receive messages from people in any groups you join. Although you normally 'subscribe' to a newsgroup, this is quite free and you can stop messages arriving by unsubscribing.

You can send links and web pages directly from within Internet Explorer, although the method varies between versions. In IE7, the toolbar has a pull-down menu called **Page**. Within the menu are options to **Send Page by Email** and **Send Link by Email**. Choosing these options will open up a new email within Outlook, showing either the full web page or just a hyperlink, depending on the option selected.

join a newsgroup

1 Open a newsgroup reader such as Outlook Express and go to **Tools – Accounts – Add– News** or click on the link to set up newsgroups.

2 Enter a display name for yourself that you want to use.

3 In the connection window, enter a news server such as msnews/microsoft.com.

4 Click on **Finish** and you will be able to download all newsgroups available.

5 To find one on a topic of interest, type this into the **display newsgroups which contain:** box.

6 Scroll down the list and find a group of interest.

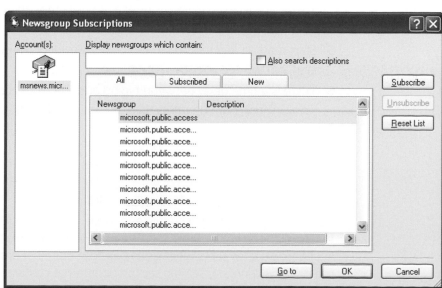

Fig. 25.20 Newsgroup1

Remember

When filling in online forms, you are responsible for the accuracy of the information you put in. A typo in your car registration data could mean that your car is uninsured.

Think about it

How many ways do you submit information online? Do you contribute to websites, wikis, or chat rooms? What other sources can you think of?

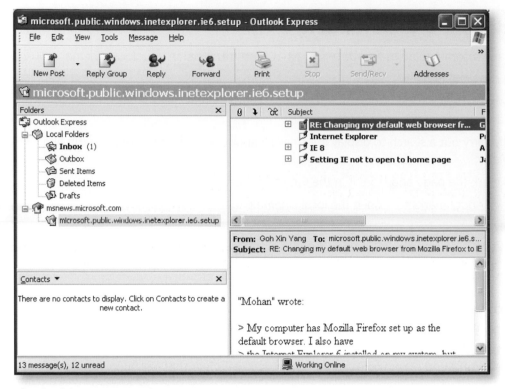

Fig. 25.21 Newsgroup2

7 You can read messages without joining by clicking the **Go to** button, or join by clicking on **Subscribe**.

8 Your Outlook Express will have the newsgroup folder added automatically and toolbar buttons will appear to allow you to reply to messages or start new posts.

Or

9 Join a newsgroup such as Google directly by going to http://groups.google.com

10 Enter keywords to locate a group of interest and click on the **Subscribe** button.

join a chatroom

1 Decide on the type of conversation you want to have and locate a suitable website by combing the word *chatroom* with keywords describing your interest.

2 Follow up any websites listed.

3 When you find one to join and decide to take part, register for free on the website.

4 You need other people online at the same time so should look for the number of current members. Many chatrooms now advertise times and dates when a chat will take place to make sure you do have other people to talk to.

5 Start reading and replying in the boxes provided. Members currently online will be listed, and you will see your own username.

6 There is a time delay between typing and the words being on display, so some conversations can be rather stilted.

Collaborative working

One of the advantages of recent advances in IT is that people can now work together despite being many miles apart. There are a number of companies that now specialise in collaborative working, but the best known is still Google.

Collaborative working encompasses a range of technologies, from shared online workspaces to Google docs, which allow users to work on documents from anywhere around the world. Real-time updates show when a file has been edited or added to. For more information on collaborative working, see the section on Internet conferencing, page 141.

Collaborative working will become more important as the working world begins to embrace cloud computing.

Check your understanding 6

1 Join a newsgroup and read some of the messages. If you want to, reply or start a new post.

2 Take part in a chatroom conversation.

Shareware and freeware

Many useful programs such as image editors, pdf readers or virus checkers are created and made available by others using the Internet. They may be completely free (**freeware**), or available at a low cost or for a limited time only (**shareware**). Where you are eventually expected to pay, you often find the program has been cut back so you can only use certain parts or see demos, and you will often find that extras you do not want such as Google toolbars will be downloaded at the same time unless you make sure you opt out.

download freeware

1 Search the Web for a free program such as the image editing program **Irfan View, Free Image Editor, Photo Scape or Paint.NET**.

2 Install it from a reputable site such as www. Cnet.com or www.ComputerActive.co.uk.

3 Click on the **Download** button and opt to save the installation program to your computer unless specifically directed to run it (**Open**) directly. Choose a convenient, temporary place to locate it, such as your desktop.

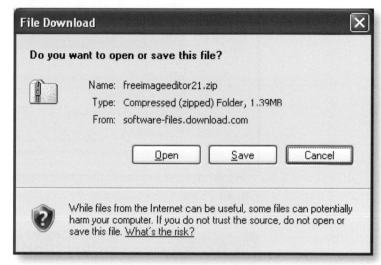

Fig. 25.22 Download

4 A window will show how long the download will take.

5 Locate and start the set up and accept the default settings. If the program files were compressed, you will first have to open an archive folder.

6 You may be offered the chance to add an icon for the program to your desktop or it will be available from your **All Programs List**.

7 If you decide you do not want to keep the program, remove it by going to **Start – Control Panel – Add or Remove Programs**, selecting it from the list and clicking on **Remove**. There may also be an **Uninstall** utility downloaded with the program available from the **All Programs List**.

Fig. 25.23 Download progress

Fig. 25.24 Installed

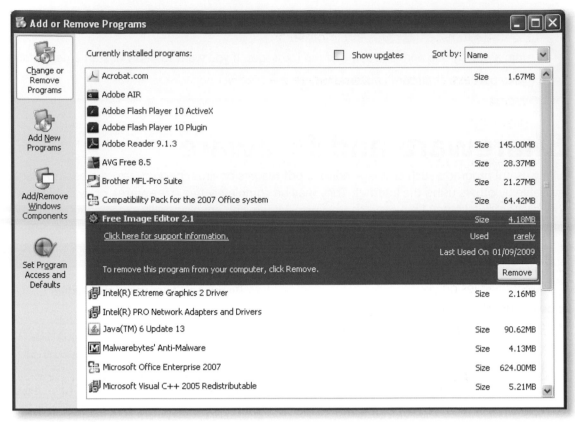

Fig. 25.25 Remove program

Check your understanding 7

1 Locate and install any free virus checker such as **AVG** or **Avast**.

2 Remove it if it is unwanted.

FTP software

The process of transferring files from one computer to another across the Internet is known as **file transfer** and it uses **File Transfer Protocol (FTP)**.

To upload your own files, you will need an **FTP program (ftp client)** and details of the correct web server – the computer where you will send your files. Common uses for such software include publishing photos to websites to share them with friends or family, using your ISP-provided web space to keep copies of important files or placing files on your own web page.

You can use your browser to upload files directly. Some sites offer their own, customised versions of FTP programs such as **Flickr** or **Mozy**, or you can transfer files within a web creation program such as **Front Page** or **Dreamweaver**.

There are also general programs available such as **SmartFTP, WS_FTP** or **CuteFTP**. When many of these programs run, you see two panes: on one side are the files on your computer and on the other the files that have been transferred. You transfer files by double clicking or dragging them across from one pane to the other, and there may be a **Restore** option once files have been uploaded.

Fig. 25.26 ftp

use FTP

1 Make sure you have the following information:
   - site name
   - host address or name/IP address where the files will be stored
   - userID
   - password.

2 Open your browser and enter the following in the address box: ftp://username@ domainname.com .

3 Enter your password.

4 Follow the instructions to transfer files.

   Or

5 Open a dedicated FTP program and click on the connect button.

6 Log in.

7 Select the files on your computer that you want to upload and, if necessary, double click to transfer them across.

8 For a website, the main page is usually named **index.html**.

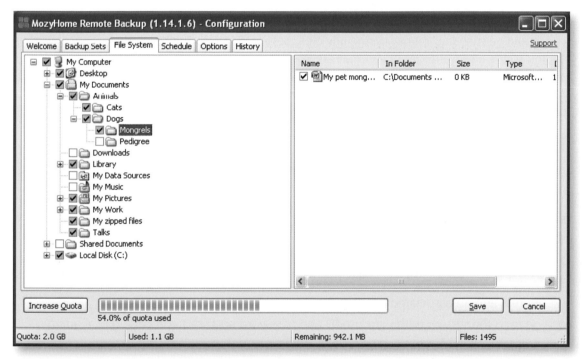

Fig. 25.27 ftp backup

Check your understanding 8

1 Download an FTP program.

2 If you have access details, upload one or more files to your web space.

Telnet

This is a program you can install on your computer that will allow you to access other networked computers remotely. It means you can alter certain aspects of another computer such as changing settings or viewing, moving or deleting files without physically being in front of the actual machine.

For it to work, you need to have the address of the remote computer, the telnet program, permission to log on and details of how to log in.

Email

There are two ways you can send and receive electronic messages:

● using software such as Microsoft Outlook 2007 installed on your computer (covered in Unit 031)

● accessing your messages online from any computer connected to the Web.

It is very useful to be able to use email without even having to own your own computer and you can do this by using **webmail** (web-based email services). Many websites offer free mail services to users, including www.google.co.uk, www.yahoo.co.uk, www.hotmail.com, www.gmx. com and www.postmaster.co.uk. Once you have an address, you can send or receive messages wherever you are in the world, as long as you have access to a computer linked to the Internet.

In the old days of dial-up, it was expensive to log on and read or send messages in this way as you would be spending money during any email session. Today, broadband services are always on and so the disadvantages of webmail are less. However, one value of webmail over an address linked to your ISP is that, if you change your ISP, your email address will not change. You can even configure software such as Outlook Express or Outlook to read and send messages using a webmail account.

Using webmail

At Level 1, you learned how to sign up for a webmail account and how to send and receive messages. The following is a summary of what you need to do.

sign up for webmail

1 Visit the website of a webmail service.

2 Click on the **sign up** button.

3 Complete the form that will include your userID becoming the first part of your email address and an associated password.

4 Log in to the system with your userID and password and start composing or reading messages.

send and receive messages

1 Click on the **Create** or **Compose** button.

2 Complete the main boxes with the full email address of people you are writing to (in the **To:** box) or copying the message to (in the **Cc:** box) as well as the message subject. Then write the message in the main window.

3 Click on **Send** to send the message.

4 Whenever you log on or click on **Check Mail**, any messages waiting for you will be received and stored in your Inbox.

5 Click on the box and double click on any message to read it in full.

6 When a message arrives in your Inbox, you can reply to it by clicking the **Reply** button, or forward it to a third person by clicking the **Forward** button. Complete any empty boxes and then send it as normal.

7 Note that **Reply All** will send your reply to anyone receiving a copy of the original message as well as the author.

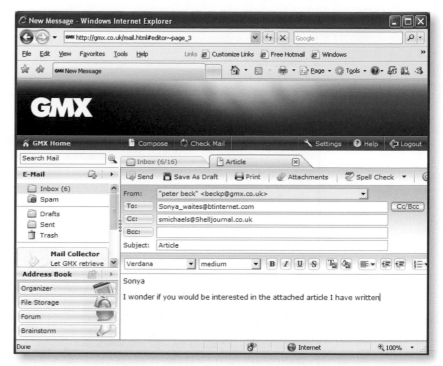

Fig. 25.28 Compose

Check your understanding 9

1 Set up a webmail account.

2 Sign in and familiarise yourself with the system.

3 Compose a message and send it to a friend or colleague who has agreed to reply. Give the message the subject *Astrology* and tell them there will be an eclipse next week. Ask them to confirm that they will come over to watch it with you.

4 Read the reply when it arrives in your Inbox.

Using file management

Once you start building up numbers of incoming messages, you may want to group them so that they are easier to work with. You can do this by creating folders in which to store the messages, and then move each message into the appropriate folder.

create folders

1 Right click on the parent folder such as the **Inbox**.

2 Select **New folder**.

3 Enter a name for the folder.

4 To create a subfolder inside new folders, right click on the new folder and create another folder inside.

Fig. 25.29 Create folder

move messages

1 Open the folder containing the messages you want to move.
2 Right click on a message and select **Move to**.
3 A list of folders will appear so click on the folder in which you want to store the message.
4 It will be moved to that folder.

Fig. 25.30 Move message

delete messages

1 Right click on an unwanted message.
2 Select **Delete**.
3 The message will be moved to your **Trash** or **Deleted Items** folder.

save messages you are writing

1 Click on **Save as Draft**.
2 The message will be retained in the **Drafts** folder until you are ready to continue writing and sending it.

print messages

1 Right click on a message.
2 Select **Print**.
3 Check the preview if it is offered.
4 Complete the **Print** dialog box options, such as number of copies, and then print in the normal way.

Check your understanding 10

1 Create a new Inbox folder named *Sky*.

2 Move into this folder the reply you received to your *Astrology* message.

3 Now delete this message and check that it has been moved to your Trash folder.

4 Move it back into the Inbox.

5 Print one copy of the message.

Automatic replies

If you are going to be away from your desk and want to let people know when you will be back, or who to contact in the meantime, you can set up your email system to send out automated replies. The following is how the system operates in GMX so your own system may differ.

Note that only the author of the message will receive the automated message, not people listed in the Cc: box.

send automated replies

1 Click on **Settings**.

2 Click on **Autoresponder**.

3 Type in the text you want in the main message.

4 Set the system to run all the time or until a particular expiry date.

5 Save the details.

Fig. 25.31 Automatic reply

Check your understanding 11

1 Check how to create an automated message in your own webmail system if it is not GMX.

2 Set up an automated reply to expire in a day or so.

3 Test that it works by asking someone to write to you within the time period set and check that they receive the automated reply.

Contacts

As you know, writing emails is simple as you can store and retrieve addresses listed in the address book. Not only can individual addresses be stored, but people who regularly receive the same message can be collected together into a grouped email address. Adding this to the **To:** box sends the message to everyone listed.

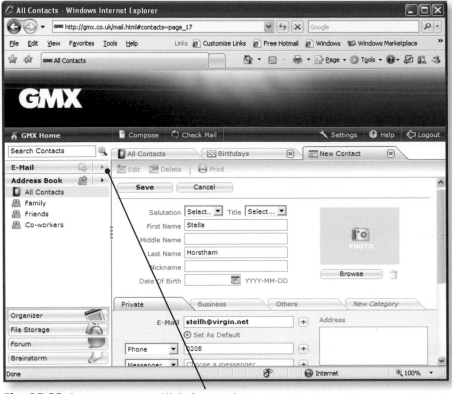

Fig. 25.32 Contact   Click for options

add a contact

1 Right click on the name in the **From:** box on an incoming message.

2 Click on **Save Address to Address Book**.

   Or

3 Click on the **Address Book** tab in the left pane.

4 Click on the right-facing arrow and select **New Contact**.

   Or

5 Click on **All Contacts** to list them in the main window.

6 Click on the **New Contact** button.

7 Enter the name, display name (nickname), email address and any other information you want to retain about them.

8 Click on **Save** to retain the details.

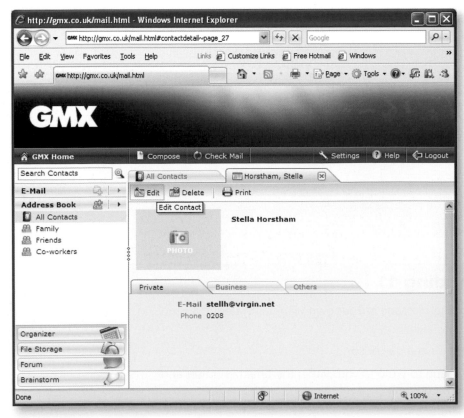

Fig. 25.33 Edit contact

edit or delete a contact

1 Click on the **Address Book** tab.

2 Click on **All Contacts** to list everyone in the address book.

3 Select the name and click on the appropriate button or first double click on the name of the contact you want to edit.

4 Click on **Edit** to display their details and make any changes before clicking **Save**.

5 To remove a contact, click on the **Delete** button.

create a group

1 Click on the **New Group** button or the right-facing arrow next to the **Address Book** tab and select **New Group**.

2 Some groups such as **Family and Friends** have already been created for you but you can now name your new group and it will be added to the list.

3 To add a new member, right click on the group name and click on **New Contact**.

4 Complete the boxes and repeat for other members.

5 To select members already in your address book, open **All Contacts** and drag their names onto the group.

6 To remove a group member, click on the group name, select the member and click on **Remove from Group**.

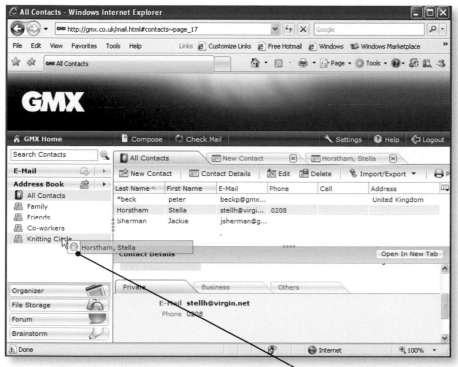

Fig. 25.34 Add to group   Adding contact to group

add contacts to a message

1 Click on the **To:** or **Cc:** box to open your address book.

2 Double click on the contact name listed in the window and it will be added to the **To** or **Cc:** box.

3 You can also start entering the email address in the message window. All contacts beginning with the same letter of the alphabet will be displayed and you can click on the correct name to add it to the box.

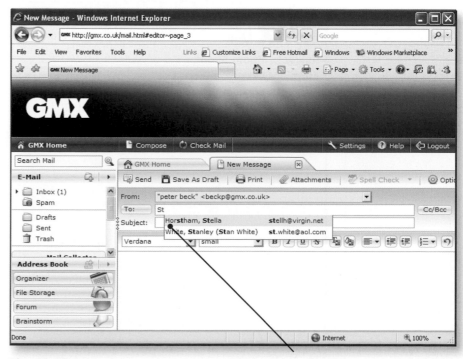

Fig. 25.35 Select recipient   Click correct name

Spam

Rules built into the webmail system mean that messages from certain domain names or containing particular words or phrases are recognised as 'spam' or junk mail that is likely to contain offensive or advertising material and so will be unwanted. It will be moved automatically to a folder already set up for you. You will have time to check that no genuine messages have been moved there by mistake before the folder is emptied.

As many unwanted messages will slip through, you can manually add these to the list of senders whose messages will be blocked and their future messages will be moved directly to the junk mail folder.

identify spam

1 Right click on the message in the Inbox.

2 Select **Mark as Spam**.

3 That message and subsequent emails from the same source will be moved to your Spam folder.

Check your understanding 12

1 Check the contents of your spam folder.

2 Right click on the folder, select **Properties** and check how many days are set to elapse before the folder will be emptied automatically.

3 Add a new contact to your address book: *Mavis Polard,* email polardm@wainfleet.co.uk.

4 Create a group named *Photo Club.*

5 Add Mavis Polard's details to this group from your Contacts.

6 Add a new member: *Stanley White,* email st.white@aol.com.

7 Create a message with the subject *Photo Shoot* to be sent to members of the Photo Club.

8 Add the message *See you all next Saturday.*

9 Save this message in your Drafts folder.

10 Print a copy of the message.

11 Delete the message.

12 Finally, take a screen print of your Contacts showing the Photo Club group listed.

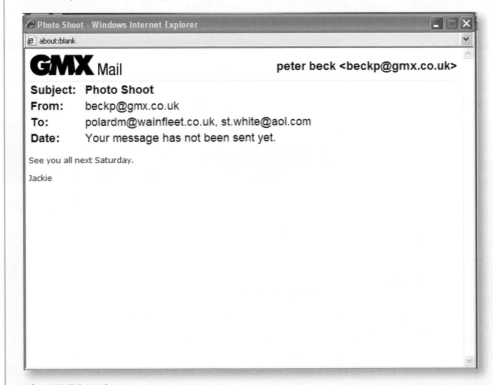

Fig. 25.36 Write to group

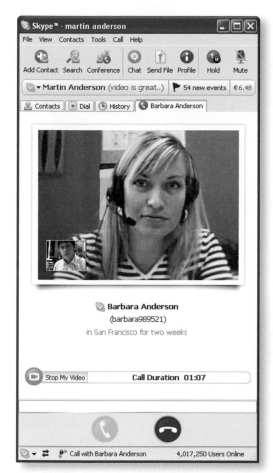

Fig. 25.37

A Skype videocall

Internet conferencing

Internet conferencing is when computer systems are used to let one or more people have a meeting without the need to travel. This technology allows them to see and hear each other as well as providing other tools to help communicate.

There are many very powerful advantages to internet conferencing, particularly the savings in time and travel costs as well as a very low impact on the environment.

The cost of just one air flight to another country with associated hotel bills would be enough to set up the facilities needed for internet conferencing.

Some of the best internet conferences can be very realistic and it's almost as though everybody is actually in the same room, especially if there's a good, fast internet connection and large screens are used.

In its most basic form, internet conferencing uses cameras, microphones and loudspeakers connected to computers to let people see and hear each other, wherever in the world they might be. This is similar to the technology used by people making Skype calls.

Skype

Skype is software that allows users to make voice or video calls using the internet to connect them rather than the traditional telephone system. Calls to other users of the service are free.

Check your understanding 13

Use the internet to visit the www.Skype.com website.

1 Visit the *Shop page* and use the *Pick a subscription* button. What is the monthly cost of unlimited world calls?

2 What limits are placed on these calls by Skype's fair usage policy?

3 Visit the *Shop* page and use the *Shop for accessories* button. Look at their Freetalk USB headset with microphone product. List what you think are the best features of this product

4 Can you find a USB headset with microphone product from another supplier that will allow Skype calls? How do you think this compares with the Skype product?

Internet conferencing tools

Fig. 25.38

Internet conferencing toolbar

Full internet conferencing also adds writing and whiteboard tools to help business people communicate quickly, effectively and professionally.

Whiteboard tools are often located in a toolbox and are used to produce diagrams or to add to photographs, plans, diagrams, maps or other images.

- The *pen* tool allows you to draw anywhere within the image. Often, selecting a pen tool shows more options such as changing the line colour or size.

- The *highlighter* tool emphasizes wording or images. Usually there is choice over the highlighter colour and size.

- *Text* tools can be used to add text to the shared work.

- The *comment* tool can be used to start a discussion or record a voice note.

- The *shape* tool is to draw lines, circles and rectangles with control over colour and size.

- The *delete* tool can remove parts of the work.

Writing tools allow people in the conference to choose to share anything they type with others in the meeting. These words can also be changed and added to by others in the meeting.

Whiteboard tools let people sketch onto diagrams, photographs, maps and other graphics to help share their thoughts and understandings. Others in the meeting may also add to these graphics.

Internet conferences can be recorded so attendees can have copies of the meeting to replay later to themselves and others.

These technologies allow people to attend meetings without leaving their workplace desk, simply using their desktop computer or laptop.

Many people use a headset with microphone so they can talk to others and hear what is said in the meeting. Others prefer the freedom of using loudspeakers and a desk microphone or a laptop with these already built-in.

A webcam can be used so they can be seen by other participants. Many laptops have webcams built into their screen surround.

Internet conferencing rooms

A lot of companies dedicate rooms in their offices as internet conferencing facilities. Typically such a room will have one or more big screens to help provide a more realistic feeling to the meetings. If the people on the screen are close to life-size it adds to the communication as it seems as though they are actually in the room.

A dedicated internet conferencing room may use large television screens connected to computer systems with sound from the TV's built-in loud speakers.

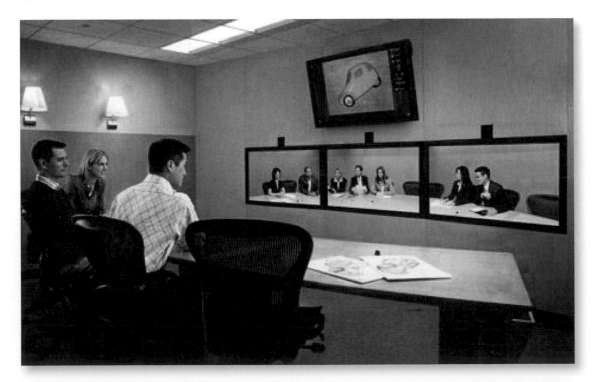

An internet conferencing room will have microphone(s) so when people speak, their voices can be heard by others in the other sites participating in the meeting.

There are companies that rent out dedicated internet conferencing rooms which are already equipped with all the equipment needed to connect to others in the meeting as well as providing professional office surroundings.

Rented dedicated internet conferencing rooms can be useful, but the technology is so affordable nowadays that most companies prefer the convenience of hosting these meetings on-site. Apart from the expense of renting conferencing facilities there is also a time (and therefore cost) overhead incurred if staff need to leave the building for meetings.

Internet conferencing connections

There needs to be a good internet connection with enough bandwidth for the live television-quality pictures to be sent and received with an acceptable number of frames per second (FPS) and without blocking or pixilation.

The more frames per second, the smoother the movement. Most would accept 15 FPS as the minimum standard for live video, with 30 FPS as a good standard.

The screen resolution is the other important factor regarding picture quality, some systems are as low as 176 x 144, with 320 x 240 for a lot of mid-quality systems. Ideally, screen resolution will be 1280 x 800 or better.

To achieve video of this quality, the internet needs to be through broadband or an even more powerful connection.

Internet conferencing software products

There are many internet conferencing software solutions available, ranging from the free to the professional (and more expensive) solutions.

Here are some examples of internet conferencing software products comparing the following features:

- **Streaming feeds number:** how many cameras can be connected to the system at the same time
- **Text-chat:** whether participants can send typed messages to others
- **File transfer:** whether you can send files to other people in meetings
- **Session Recording:** whether you can record a meeting's video and audio
- **Web-based:** if no, you need to download software to use the service

	Streaming Feeds Number ⬍	Text-Chat ⬍	File Transfer ⬍	Session Recording ⬍	Web-Based ⬍
Adobe Connect Now	3	Yes	Yes	No	Yes
EkkoTV	3	Yes	No	No	Yes
FlashMeeting	No Limit	Yes	No	Yes	Yes
iVisit	8	Yes	Yes	No	No
MeBeam	16	Yes	No	No	Yes
MegaMeeting	16*	Yes	No	No	Yes
Oovoo	6*	Yes	Yes	Yes	No
PalBee	9	Yes	Yes	Yes	Yes
SightSpeed	9*	Yes	Yes	Yes	No
TokBox	20	Yes	No	No	Yes
Vawker	N/A	Yes	No	No	Yes
Vidivic	9	Yes	No	No	Yes
VSee	No Limit	Yes	Yes	Yes	No
WengoMeeting	5	No	No	No	Yes

Fig. 25.39 Comparing Internet conferencing software

Adobe Connect Now

Adobe Connect Now can create online meetings with video-conferencing, VoIP conversations, whiteboard, shared files, chat and you can share your screen. It is free to use after registration.

http://www.acrobat.com/

EkkoTV

EkkoTV is a Flash-based service that enables video conferences and is free to use. Even without registering, you can enter your details for a new room to be created. The URL is then needed for the other participants to attend.

http://www.ekko.tv/

FlashMeeting

FlashMeeting is a video conferencing application with co-browsing. It uses an Adobe Flash 'plug in' and Flash Media Server and then works with your web browser. The leader needs to register and book meetings. Participants use the URL to attend.

http://flashmeeting.open.ac.uk/

iVisit

iVisit is free for mobile and desktop computers enabling video conferences with up to 8 people who can share hi-res photos, text-chat, and track mobile users with an integrated GPS system. This software has professional upgrade plans available.

http://www.ivisit.com/

MeBeam

MeBeam is a free Flash-based video conferencing tool that does not need to have anything installed on your computer. No registration is needed; a video conference is started by entering its name, then shared with attendees.

http://www.mebeam.com/

MegaMeeting

MegaMeeting is a web based video conferencing solution using your internet browser with broadband internet connection. Up to 16 individuals can participate in a meeting, with an unlimited number of additional secure attendees able to see and hear the meeting.

http://www.megameeting.com/

Features of internet conferencing packages

Feature	What it does
Application sharing	Share active applications such as Photoshop, Auto-CAD, or others whilst in a meeting to collaborate and communicate.
Audio / video messaging	If an attendee isn't available, this feature can send them an audio or video message they can see later.
Authoring information	This feature can use content from meetings to help create and maintain websites that distribute information to team members and can connect members with queries to those who have the answers.
Background noise suppression	This helps to make the speaker's voice a lot easier to hear because other noises in the room are reduced before the sound is sent to the attendees.
Brainstorm templates	This tool allows ideas to be written down and linked together as they get suggested by attendees. Usually these can be moved around and re-arranged as the ideas and thinking evolves.
Central management	Central management allows one person to lead and control the internet conference.
Chat	Similar to instant messaging such as MSN, chat make communication even easier one-on-one or with all the meeting participants
Co-browsing	Show your browser exploring the web, so others can see the sites you visit whilst in a live meeting
Collaborative tools	These tools help people in the conference work together such as discussions, file sharing and many others.
Cross platform	Allows internet conferencing with any user on their desktop, laptop, or mobile phone.
Desktop sharing	Allows participants see what you see on your computer
Document sharing	Internet conference users often want to share documents live. Most internet conference packages are able to share Word and PDFs documents and the better ones allow other types as well.
Echo cancellation	When using sound, an echo can be intrusive and annoying because your words are picked up by another attendee's microphone and sent back to your loudspeakers. The effect of this is that you hear your words again, like an echo. Echo cancellation stops this.
Firewall traversal	Firewalls are on every professional internet connection so that attacks from viruses and other internet threats are reduced or eliminated. Some internet conferencing packages offer firewall traversal to enable fast and secure communication with other sites in the meeting.
Flash-based presentation tools	Microsoft PowerPoint is very commonly used by people to produce slides supporting their presentations. The documents PowerPoint produces (PPT) are surprisingly difficult to use for internet conferencing as they do not work well in browsers. Flash-based presentation tools can convert PPT presentations to give web slides additional animation, sounds, and motion.

Free web conferencing software	There are many promises of free web conferencing software solutions, which are designed to lure user into paying for costly monthly services that are free only during a trial period. Other free web conferencing software may use a beta version, so the presentation may not be as smooth as a fully tested solution because the software designers are still trying to work out all of its bugs and test reliability. Free software can be a reduced version of the premium solution, not including extra features, so users are tempted to pay to upgrade later. Some web conferencing tools are free because the authors want to increase their web conferencing visibility among other users of their primary business.
GPS viewer	On some internet conferencing systems, mobile users can broadcast their GPS position on a map which can be seen by attendees.
Multimedia conferencing	This is what most people expect from their internet conferencing system, allowing meetings with a mixture of sound, video and animations.
Multiparty meetings	Where more than two participants can meet using internet conferencing to share video, presentations, documents, applications, browsers, and desktops.
Online chat rooms	Some internet conferencing systems allow several online chat rooms, so people attending the meeting can type messages to each other which are not seen by the rest of the conference.
Polling	This is a facility where attendees can vote on a topic, with the system counting the votes and show the result.
Presentation sharing	This is an internet conferencing system that can share PowerPoint presentations while in a live video conference by supporting the .PPT and .PPTX formats used to save PowerPoint documents.
Recording	Save, replay, post or email valuable interactions and presentations – including audio.
Remote PC access	Most internet conferencing system allow remote PC access so attendees can use their computers from anywhere with a good internet connection to join the meeting.
Share keyboard and mouse control	Securely collaborate with colleagues on projects by working together in real time on the same application.
Specific application sharing	Share only the application you choose, keeping attendees focused on what you want them to see
Sync / unsync modes	Sync is synchronised whilst unsync is unsynchronised, referring to how the screens in a meeting co-ordinate. Sync mode forces synchronization of pages, so when pages are changed, all other user screens will also change pages. This means that everyone always sees the same pages at the same time.

ooVoo

ooVoo is video-conferencing software in beta form, so it is free to download, use and is available for both Windows and Mac computers. After registering, you can communicate with up to six people through text-chat, video-audio conference, record video messages and share up to 20 files at once, up to 25 MB per file.

http://www.oovoo.com/

PalBee.com

PalBee.com is a newly launched video conferencing system that is completely free to use for up to 10 people, with whiteboard, PowerPoint presentation uploading and up to an hour's recording available.

http://www.palbee.com/

SightSpeed

SightSpeed is a cross-platform video conferencing system that works on a less powerful computer, is quick and cheap. It allows video calls with up to 9 people, text-chat, file sharing, session recording and sending video messages. Free for two people, or $9.95/month for 4 people and unlimited video storage. A new web-based version of the program is also available.

http://www.sightspeed.com/

TokBox

TokBox is a free web-based video conferencing application allowing online video meetings with 20 or more people. After registration, you create a video room where you can invite participants for a video conference. This can be embedded on your site, or login in to your TokBox page to set attendee emails for the conference.

http://www.tokbox.com/

Vawkr

Vawkr allows you to get your own video chat room. After you register and create the room, you can invite other people to join by sending them your room's URL. The simple controls are just for volume and microphone. The service is completely web-based and free.

http://vawkr.com/

Vidivic

Vidivic allows video conferencing using a webcam and PC connected to Internet. After free registration, your can create meetings up to 4 hours with 9 participants at once, to see, talk and chat with each other. Currently in beta form, it is completely free to sign up and use.

http://www.vidivic.com/

VSee

VSee is a free videoconferencing and application sharing service. You can remotely edit and annotate documents, share applications and desktops, transfer files, record and share videos, pan, tilt, and zoom remote cameras. VSee is free to use for an unlimited number of people.

http://vsee.com/

WengoMeeting

WengoMeeting is a free flash web-based service allowing up to 5 user video conferencing meetings without software downloads. After registering, add participants' email addresses and a web conferencing room is created.

http://www.wengomeeting.com/index.php

Sync / unsync modes (contd.)	When sync is on viewers cannot navigate pages at all. They are bound to follow you around from page to page. When sync mode is off viewers can change their pages as they wish. During a presentation, sync would be on so that everyone follows from page to page.
URL	The Uniform Resource Locator is the technical name for a web address such as http://www.pearson.com
Video conferencing	Uses cameras, microphones and speakers so attendees can see and hear each other.
VoIP	Voice over the IP (Internet Protocol) is a technology that allows people to make "phone calls" using their computers and broadband. Usually this is much cheaper than phone calls made through the telephone network as broadband has a flat monthly fee, rather than paid on a charge per call basis.
Voting	This feature is also known as polling, allowing attendees to vote on an issue with the results collated and shown.
Web browser plug-in	A web browser plug-in is software that works with web browsers, such as Internet Explorer or Firefox, adding extra capabilities to the browser. Web browser plug-ins for internet conferencing allowing participants to use their usual web browser to attend the conference.

Check your understanding 14

1 How many internet conferencing software solutions can you find?

2 What features are missing in the cheaper solutions?

How to access a conference program

There are a range of conference programs available, and the features offered are largely similar. Here we use the Vsee conferencing system as an example.

To use Vsee you must first visit their website www.Vsee.com to download a copy of the software to your computer. This is free to personal users or available as a free trial for businesses (other programs will vary).

Once the software is downloaded and installed you can run it to access this conference program and to register an ID for Vsee.

Starting a conference call

Start the Vsee program then choose **Host a meeting** from the main menu. You can now see a list of the other Vsee people you know from your address book and add them to your meeting.

Receiving a conference call

Receiving a conference call is automatic if Vsee is running when the other person calls you.

Transferring files to a connected user

To transfer a file with Vsee, you drag the file onto their image and drop it. They can then drag the file from their image on their computer onto the desktop or into a folder.

Using a whiteboard facility

With Vsee the whiteboard is any program you can use on your computer which is then shown to the other attendees in the conference. You can select **Share applications** from the main menu to allow others to contribute.

Exiting a call correctly

Say good bye to the other attendees then close your conferencing program of choice.

Sharing a desktop

Some conferencing programs allow you to share your desktop with other members of the group. This works in a similar way to whiteboard facilities (above), and enables members of the same conference to view applications, presentations, and documents as if they were hosted on their own computer.

Security over the Internet

There are various activities that can take place over the Internet that you need to guard against. These include accidentally downloading harmful viruses, having your personal information accessed or becoming a target for fraudsters and criminals.

Although you will be aware of the use of passwords and userID data, these will only allow you to access your own secure areas on networks or the Internet, and will not help you avoid many of these other dangers. Fortunately there are ways to set up your computer to lessen the chance of these activities being successful.

Did you know

Your online passwords are as important as your PIN. Do not share your access details with anyone – at best, your social networking accounts will be compromised. At worst, you could become a victim of identity theft.

Remember

Online security is not just about protecting your data or your money – you must also protect yourself. Do not meet online friends without putting basic safety precautions in place: bring a friend, meet in a public place, and don't put yourself in potentially risky situations.

Remember

There are a lot of things to think about when it comes to staying safe online: check firewall and internet security settings; don't disclose personal details or other inappropriate information; and report any suspected security threats, from malware to phishing emails.

Virus protection

All computers should run anti-virus programs that will either prevent these harmful programs being downloaded or will identify and neutralise any that get through. As new viruses are being written and sent out over the Internet all the time, it is very important to keep your virus checking software up-to-date.

change virus protection settings

1 Open the anti-virus program that you are running. You may have an icon on the desktop or in the taskbar tray.

2 Depending on the program, the controls will look different, but the following screen shows the AVG user interface. The various components will be visible.

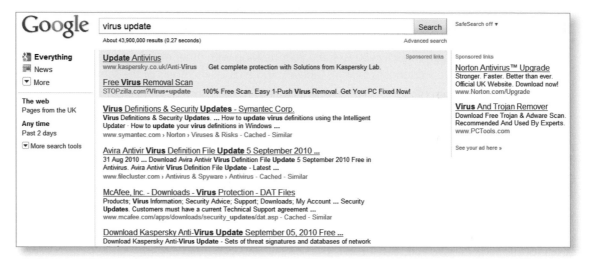

Fig. 25.40 Virus update

3 The two settings you will want to control are updating the virus database and scanning your computer on a regular basis.

4 To set these activities, go to **Tools – Advanced Settings**.

5 If you click on **Update** in the left-hand pane you can change when to update the files – for example, on computer restart.

6 If you click on **Schedules** you can set the virus database update to take place at the same time each day or even several times a day, and when to scan the whole computer.

Internet locking software

There are now many programs available that you can use to monitor and lock access to your computer through the use of passwords and encryption facilities. These include, for example, **Internet Lock**, **WinGuard** and **Desksense**.

Spyware and adware

As well as programs that may damage your computer, other unwanted files may be sent to you when you are browsing. These include **spyware** that tries to detect your browsing history or keystrokes and send it to third parties, and **adware** that sends advertising material in the form of annoying pop-up windows.

There are many applications available, such as **Spybot**, **Ad-Aware**, **Super Antispyware** or **Stopzilla**, that you can install to identify and remove these programs. Like anti-virus programs, they need to be updated on a regular basis to work effectively.

Firewalls

Operating systems are not secure and so a **firewall** is a device, either software or hardware, that protects you when you are on a network such as the Internet. It enables you to decide what applications or data can be passed to or from your computer and it does this by setting up a series of rules that limit who has permission to gain access. Some normal activities such as sending and receiving emails will have permissions pre-configured, but others such as instant messaging services will need your agreement to run.

change firewall settings

1 Go to **Start – Control Panel** and open Windows Security Settings.
2 Check that the Firewall is **On**.
3 To make changes, click on the **Windows Firewall link**.
4 For example, click on **Exceptions** if you want to enter details of any network connections and services that will be allowed through in less secure locations even if the firewall is on.

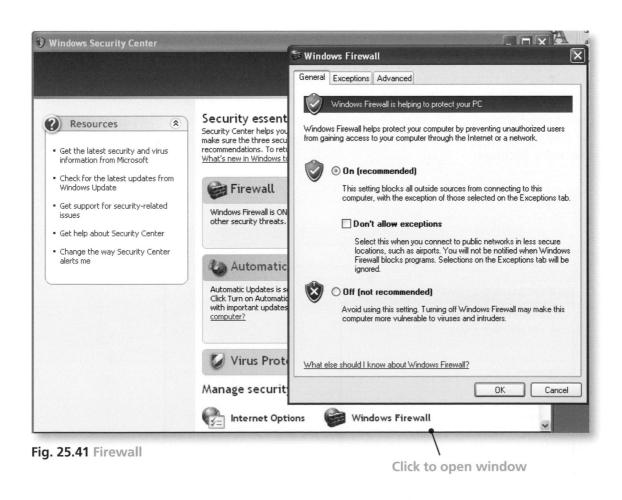

Fig. 25.41 Firewall

Click to open window

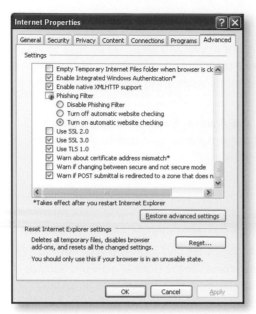

Fig. 25.42 Alerts

Alerts

So that users are aware of what is going on, some applications such as anti-virus software programs or firewalls display warning messages known as **alerts**. These will be set to appear if, for example, security features are turned off. Alerts also appear if warnings are necessary – for example, to tell you a website security certificate has expired (see below). You can also set some Windows warnings yourself.

change settings

1 Click on the **Internet Options** link in the Windows Security Centre window.

Or

2 Open your browser and go to **Tools – Internet Options** and click on the **Advanced** tab.

3 Take off or add ticks to checkboxes related to various warnings.

Digital signatures

Digital signatures are used for electronic documents such as financial or legal material to authenticate the originator or ensure that document contents have not been changed during transmission. The person receiving the message or document can then be reassured that the sender and/or the documents are genuine. Digital signatures can be time-stamped and may be used to prevent someone later denying that a document has actually been sent.

To use digital signatures, you need a digital certificate provided by an approved certificating authority that will verify your identity.

Messages or documents copied and sent via email are turned into mathematical summaries known as hashes that are then encrypted (i.e. scrambled so they read as nonsense) using a special private 'key' that only you have. This encrypted data becomes your digital signature.

Most email systems will allow you to encrypt messages. For example, Outlook 2007 (covered in Unit 031) offers this if you click on the **Message Options** command and then click on **Security Settings**.

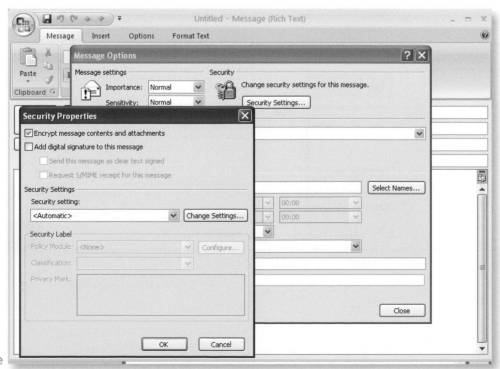

Fig. 25.43
Encrypt message

When received, the message is again hashed and a public key is used to decrypt the data. If the hashes match, the document is genuine.

Website certification

A file known as a **digital certificate** should be installed on secure Web servers to identify websites as genuine. This is to reassure visitors such as online shoppers that they can trust the website, particularly when providing information about their credit cards or other personal details.

To obtain a certificate, the companies and their websites are verified by a third party, such as VeriSign. Once legitimacy has been established, they will issue an **SSL certificate**. This digital certificate is installed on the Web server and will be viewable when a user enters a secure area of the website.

The main signs that you are visiting a secure page are the URL starting **https://** rather than http:// and a padlock icon appearing on the screen. To view the certificate, click on the **padlock**.

As certificates normally last between one and three years, it may become out-of-date before the company realises and renews it, and you may see an alert warning you of this when you visit the site. You can then decide whether to proceed with your transaction or not.

The biggest threat to your safety as an IT user is other users. Some people take advantage of the anonymity of the Internet to commit crimes, from hacking to identity theft.

Malicious programs can harm your computer, but malicious people can harm you. Do not give out personal information online, especially anything that could help people identify you in real life – for example your address or phone number, or banking details.

When sharing photos or other information on a social networking site, check your privacy settings to make sure you are only sharing them with people you have approved.

Phishing

One area where website certification can make a difference is in the fraudulent practice of enticing computer users onto a fake website where they are encouraged to provide important personal information such as their bank details or passwords. This activity is commonly known as **phishing**, and it normally takes the following form:

1 You receive an email that appears to be from your bank, asking you to re-enter your personal details as they have been lost.

2 A link inside the email is provided as a quick way to visit their website.

3 The website is in fact a fake.

4 Buttons or forms on this website direct you to enter all your personal details.

If you are suspicious of such an email:

• Do not click on any link inside the message.

• Use your normal method to access your bank website.

• Ask them to confirm what details they need. They will probably tell you that they have not made any such request.

• If you do make the mistake of clicking on the link, or you ever visit a website that makes you suspicious, check that the site certificate is up-to-date.

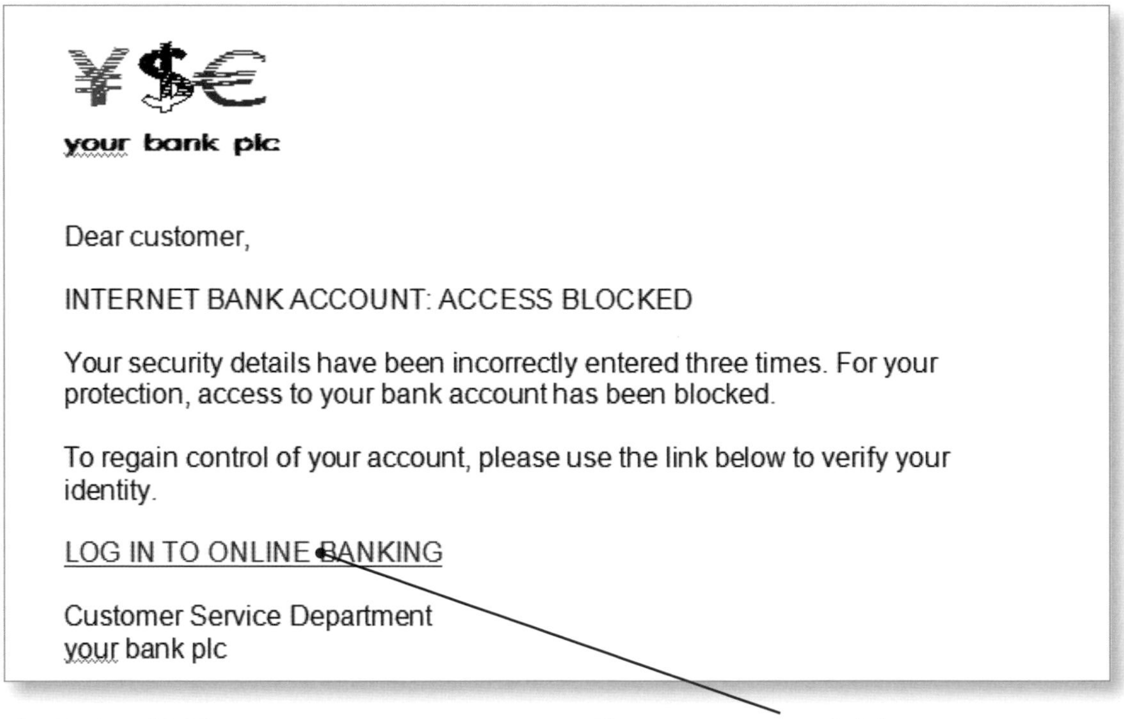

your bank plc

Dear customer,

INTERNET BANK ACCOUNT: ACCESS BLOCKED

Your security details have been incorrectly entered three times. For your protection, access to your bank account has been blocked.

To regain control of your account, please use the link below to verify your identity.

LOG IN TO ONLINE BANKING

Customer Service Department
your bank plc

Fig. 25.44 Phishing   They want you to click here

Copyright

A different form of protection is afforded to the author of original material such as music, videos, software, pictures or written material. This is the law of **copyright**.

Although you may want to publish and advertise your products over the Web, you will not want others making money out of your artwork. The 1988 Copyright, Designs and Patents Act gives you the right to control how your material is used and to be identified as the author. This right lasts between 25 and 70 years, depending on the type of material you have produced. For example, copyright on songs lasts for 50 years after a recording is first released, and for a film or a literary, dramatic or artistic piece of work copyright lasts for 70 years after the death of the author.

You may have seen the copyright symbol © next to a named piece of work, but you have copyright protection even if this symbol is not shown.

Check your understanding 14

1 What is phishing and what can you do to avoid it?

2 Check that your firewall is turned on.

3 What two visible signs are there to show that you are visiting a secure website?

4 Why would you install and use adware?

5 Access your virus protection software and run a scan of your computer.

Assignment

This practice assignment is made up of four tasks

- Task A – Customise a web browser and security
- Task B – Surfing the web, newsgroups and forums
- Task C – Using email
- Task D – Internet conferencing

Scenario

You work for a small computer shop, helping in all the sections. There has been recent demand from customers on information and help about using the Internet so the owner has asked you to produce some pages as a free handout for the shop counter and also for download from the shop's website.

Please read the text carefully and complete the tasks in the order given.

Task A – Customise a web browser and security

1   Create a document named **Shop internet handout**. Add screen shots to this document as you complete the tasks in this practice assignment.

2   Customise your browser with your choice of site as the home page.

Add a screen shot of settings to do this to **Shop internet handout** document.

3   Add another two buttons of your choice to the browser.

Add a screen shot of settings to do this to **Shop internet handout** document.

4   Remove the command bar toolbar, take a screen shot for the **Shop internet handout** document.

Show the command bar toolbar again with another screen shot for the **Shop internet handout** document.

5   Add the **Shop internet handout** document to the section on browser options, showing how to set these with brief explanations on when you should:

   a)   Enable or disable images
   b)   Configure web page access restrictions
   c)   Enable or disable the acceptance of cookies
   d)   Enable or disable the firewall
   e)   Use virus protection software

Task B – Surfing the web, newsgroups and forums

1   Use a search engine to locate the City & Guilds e-Quals website.

Find the specification for the Level 2 'Using the Internet for finding, selecting and sending information' course (7266/7267-025) on this website.

Add screen shots of the searches to do this into your **Shop internet handout** document.

2   Add to the section on navigation in the **Shop internet handout** document explaining how to move between web pages using:

   a)   URLs
   b)   Textual and graphical hyperlinks (hotspots)
   c)   Selecting from the favourites/bookmarks menu

3   Visit the www.bbc.co.uk website.

Click on **News**, then **Have your say**.

Click on a story you find interesting then click on the **Create your membership** button.

Register with the website. Respond to the email sent from the website by clicking on the link then using the **I agree** button to confirm your registration.

Post a reply to a story on the website.

Add a screen shot of this to your **Shop internet handout** document.

4   Visit the www.avforums.com website.

Register with the website. Respond to the email from the website by clicking on the link to confirm your registration.

Find a discussion forum you find interesting and participate in it.

Add a screen shot of this to your ***Shop internet handout*** document.

5   Visit www.downloads.com to locate a shareware or freeware application you would find useful. Download and save the application to My Documents.

Add a screen shot of the download in progress to your ***Shop internet handout*** document.

6   Visit www.downloads.com then search on ***Earth WP***. This should find a wallpaper zip download named Earth WP 1 (description: View planet Earth on your desktop).

Download and save this to My Documents.

Open this compressed file to view the contents. Take a screen shot of this to your ***Shop internet handout*** document.

Extract the zipped files to My Documents.

Add a screen shot of the extraction in progress to your ***Shop internet handout*** document.

Task C – Using email

1   Create an email account using Google so you have a new Googlemail web-based email address.

Add screen shots of this to your ***Shop internet handout*** document.

2   Use your new Googlemail web-based email address to create, send and receive email messages to and from your usual email account.

Save, print and then delete these emails.

Add screen shots of these to your ***Shop internet handout*** document.

3   Create a single compressed file from a folder in My Documents. Send an email to your usual email account with the compressed file as an attachment.

Add screen shots of this to your ***Shop internet handout*** document.

4   Configure your email application to give an automated reply to new email messages saying you're out of the office until next week.

Add screen shots of this to your ***Shop internet handout*** document.

5   Use your email address book to

   a)   Add addresses
   b)   Edit existing addresses
   c)   Delete addresses
   d)   Set up a group

Add screen shots of this to your ***Shop internet handout*** document.

6   Use your email to create and send messages to an email group.

Add screen shots of this to your ***Shop internet handout*** document.

Task D – Internet conferencing

1   Visit www.VSee.com. Download their conferencing software then install it onto your computer.

Run this download and register with them to set up an ID.

Select Demo from the address book and ask them to share a whiteboard.

Add screen shots of this to your ***Shop internet handout*** document.

2   Start a conference using VSee with a friend or colleague. Transfer a document to another connected user in your conference.

Exit the conference.

Add screen shots of this to your ***Shop internet handout*** document.

Presentation software

This unit involves the use of the presentation software Microsoft PowerPoint 2007 to create complex layouts and designs. You will learn how to use a master slide to maintain continuity throughout a presentation, how to amend and control slides and how to add interest when running a slide show.

At the end of this unit you will be able to:

⊕ create, save and use a template

⊕ add text and control its attributes

⊕ add graphical objects

⊕ add animation and multimedia objects

⊕ modify entries

⊕ produce hardcopy

⊕ order and run slideshows.

Launching PowerPoint

When the program opens, you will see one slide in the main window temporarily named Presentation1 plus panes offering different options. This is known as **Normal view**.

Views   Slide/outline pane   Slide in main window

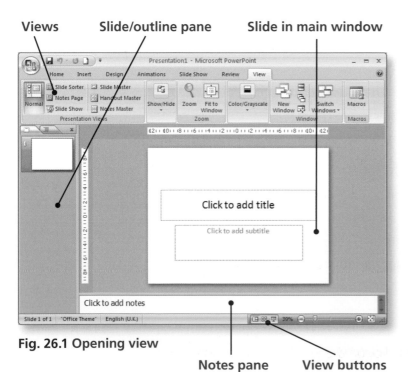

Fig. 26.1 Opening view

Notes pane   View buttons

launch PowerPoint

1 Click on the **Start** button.

2 Go to **All Programs – Microsoft Office – Microsoft Office PowerPoint 2007**.

In Normal view you can:
- work directly on a slide
- enter text onto the slide in an **Outline** pane
- view thumbnails of any slide on a **Slide** pane (which alternates with Outline pane – click on the tab to change view)
- add speaker's reminders into the **Notes** pane.

On the **View** tab or from a button at the bottom of the slide you can also move to:
- **Slide sorter view** – an alternative way to view all the slides
- **Slide show** – the view in which to run through a presentation on a computer
- **Slide Master** – this offers a way to create a template on which all the slides in the presentation will be based.

close a presentation

1   Click on the **Close** button.
    Or

2   Go to **Office Button – Close**.

3   To exit PowerPoint, go to **Office Button – Exit PowerPoint**.

Saving

As well as saving presentations as **.pptx** files, you can save them as shows, web pages, templates, images or other file types. Select the file type in the **Save as type:** box when saving.

save a presentation

1   Click on the **Save** button.

2   Select the location in the **Save in:** box.

3   Rename the file if necessary.

4   Change the file type if saving in a different format.

5   Click on **Save**.

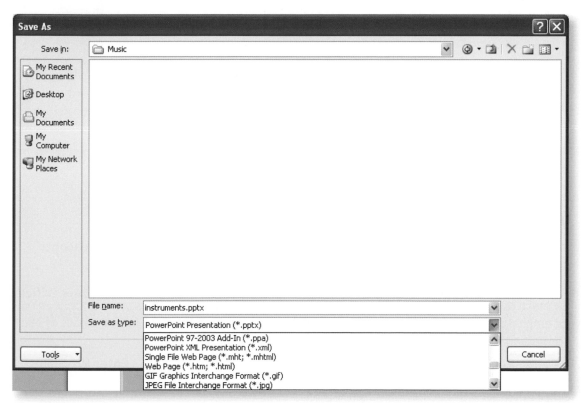

Fig. 26.2 Saving

Slide layout

Slides come with pre-set areas for entering text or other content known as **placeholders**. The first slide is a **Title slide** and has areas for a title and subtitle. When you add new slides, these will have a **Title and Content layout** unless you have previously selected a different layout, in which case they will repeat the selected style.

change slide layout

1   Right click on any slide and select **Layout** or click on the **Slide Layout** button on the **Home** tab.

2   Scroll through the selection and click on your preferred layout.

3   This will now be applied.

Slide layout

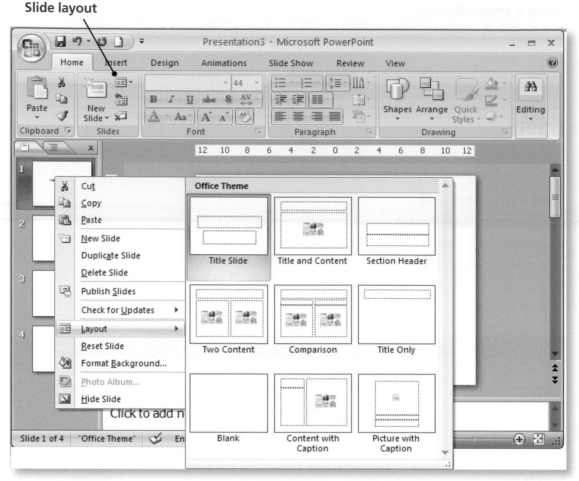

Fig. 26.3 Slide layout

Adding text

As well as typing into a text placeholder, you can add text to any part of a slide by creating a text box. In PowerPoint 2007 the default font is Calibri – size 44 for main headings, size 32 for subheadings and size 18 in a text box.

To make changes to these defaults, select the text and apply an alternative style, size, font colour or emphasis from the drop-down buttons on the **Home** tab. (See Unit 022 for full details of text formatting.)

To set a size that is not displayed, type it in over the size in the box and press **Enter** to confirm the setting.

add text into a placeholder

1 Click into the box and type your own entry.

2 To set text on a new line within the placeholder, press **Enter**.

3 If you type too much text for the box size, an **AutoFit Options** button will appear. Click on **AutoFit** to reduce the font size so that it fits the box, or click on **Stop** to leave the font size as set.

4 You can use the alignment buttons to set the text to the left or right of the placeholder.

5 Select any text and press the **Delete** key to remove unwanted entries.

Format entries

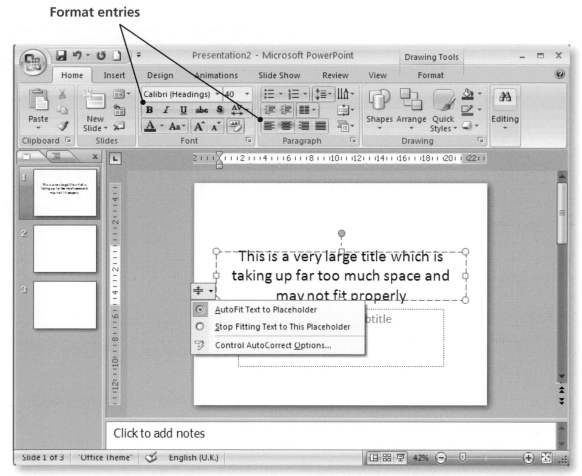

Fig. 26.4 Add text to placeholder

add text in a text box

1 Click on the **Insert** tab.
2 Click on **Text Box**.
3 Move the pointer onto the slide and drag out a box.
4 The cursor will be positioned inside and you can start entering your text.

copy in text created elsewhere

1 Select the text in the original document.
2 Right click and select **Copy**.
3 Click on the **'Click on to add...'** text inside a placeholder.
   Or
4 Click next to a slide icon on the **Outline** tab.
5 Right click and select **Paste**.
6 If there is a large block of text, you may need to select it and reduce it in size to fit on the slide.
7 If you first right click on an empty part of a slide rather than inside a placeholder, the copied text will be pasted into a newly created text box. If necessary, click to select the border and resize it by dragging.

Pasted onto blank part of slide

Pasted into a placeholder

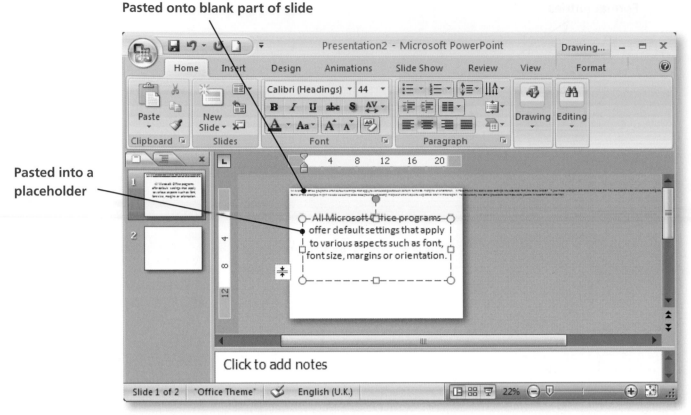

Fig. 26.5 Copied text file

Check your understanding 1

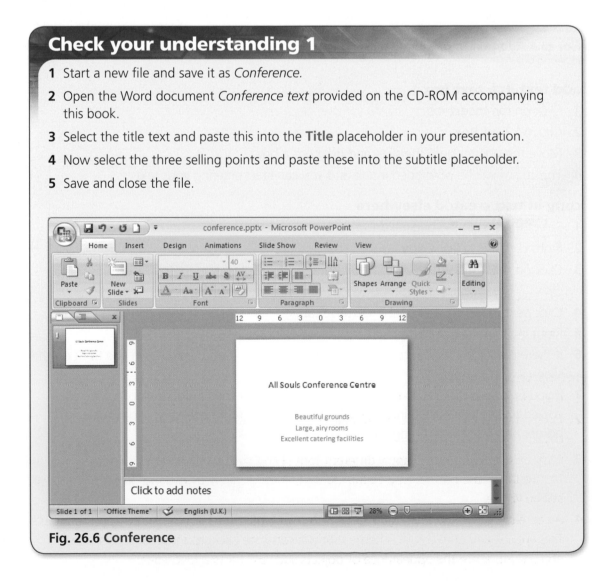

1. Start a new file and save it as *Conference*.

2. Open the Word document *Conference text* provided on the CD-ROM accompanying this book.

3. Select the title text and paste this into the **Title** placeholder in your presentation.

4. Now select the three selling points and paste these into the subtitle placeholder.

5. Save and close the file.

Fig. 26.6 Conference

Using Master slides

All Microsoft Office programs offer default settings that apply to various aspects such as font, font size, margins or orientation. In PowerPoint, the **Slide Master** stores information about the layout of each file. If you make changes here, they will be reproduced throughout the presentation.

Some of the common changes made to the master slide include colouring slide backgrounds or adding images or other objects (which are explained later in this unit). Follow exactly the same procedure but make sure you are in Master slide view first if you want all the slides to have the same appearance.

You can also insert header or footer entries or apply different font styles to headings and subheadings. You will see that these are set at various levels, each one having its own indent and style of bullet point. To change any of these, first select the relevant level.

As you may want the Title slide of a presentation to look different, there is a **Title Master** as well as **Slide Master**. To apply settings to every slide including the title slide, you must make sure you select the **Office Theme Slide Master** – the top slide in the pane – when setting up the master slide.

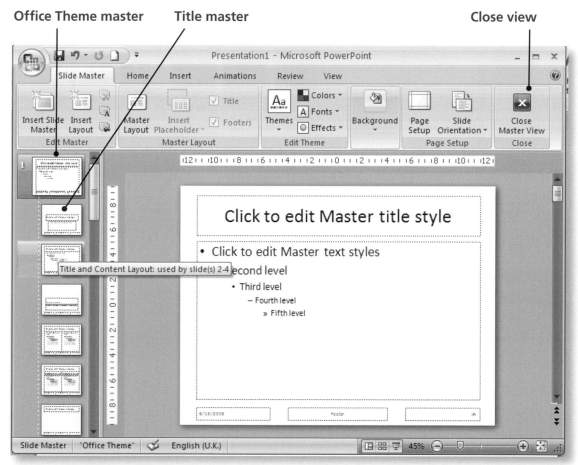

Fig. 26.7 Master

set up the master slide

1 Click on **Slide Master** on the **View** tab.

2 Click on the **Office Theme** thumbnail, or **Title Master** thumbnail to change that slide only.

3 Click on any text entry to apply different font styles, sizes, colours or bullet points from the **Home** tab in the normal way. Return to **Slide Master** view by clicking on the tab that will still be visible.

4 As an alternative, use the built-in gallery to select a theme from the button on the toolbar. This applies an entire colour scheme that will combine backgrounds, fonts and borders and will influence the appearance of objects such as charts added later.

5   If required, change orientation or margins using options from Page Setup, available on either the **Design** or **Slide Master** tab.

6   Return to **Normal view** by clicking on the **Close Master view** button or the **View** button at the bottom of the screen.

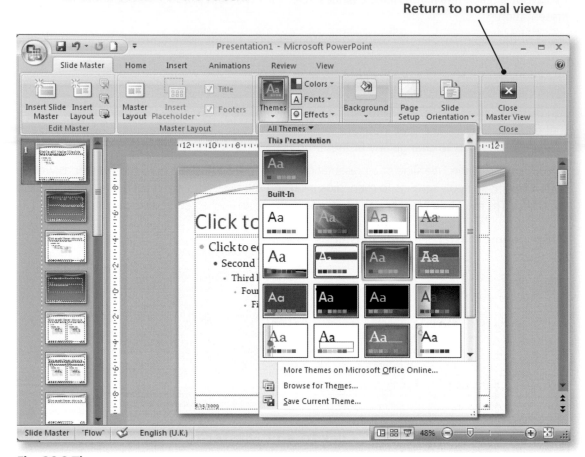

Fig. 26.8 Themes

Check your understanding 2

1   Start a new presentation.

2   In **Slide Master view**, select several different themes and see the effect.

3   Change the font type and size for the title text.

4   Change the slide orientation.

5   Return to **Normal view**.

6   Close the file without saving.

Using templates

If you want to use new designs more widely than in the current presentation, applying changes made to the master slide to future new slideshows, you can save the presentation as a template rather than a normal PowerPoint file. Your settings will now be available to use in any new presentation.

save a presentation as a template

1   Open the **Save As** dialog box.

2   Name the template.

3   Click on the **Save as type:** box and select **Template** as the file type.

4   A new Templates folder will appear in the **Save in:** box.

5   Click on **Save** to complete the save.

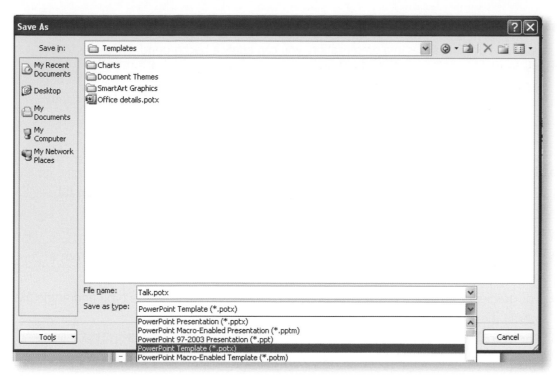

Fig. 26.9 Template

work with a template

1 Go to **Office Button – New**.

2 You can select a template from the gallery on which to base your new presentation, or locate a template you created and saved.

3 Click on the name of any template in the **Recently Used Templates** section, or click on the **My templates** link.

4 All the templates you have created will be displayed.

5 Click on one and then click on **OK**.

6 You can now make changes to any slides and save the presentation in the normal way.

View your own templates

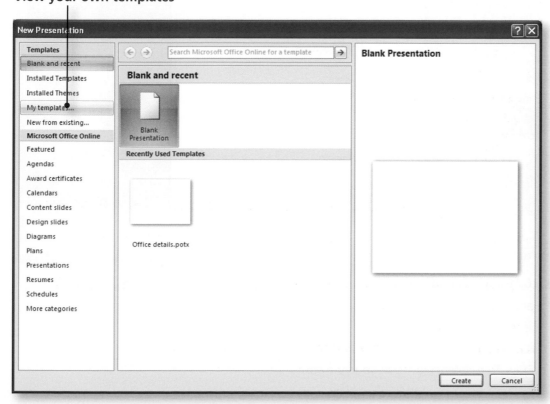

Fig. 26.10 Use template

Adding backgrounds

To colour one or more slides in a presentation, you can apply a background. There are four choices:

- solid colour
- a gradient of two or more colours
- a picture
- texture

apply a background

1 Click on the drop-down arrow in the **Background Styles** box on the **Design** tab.
2 Click on **Format background**.
3 Select a type of background by clicking in the radio button.
4 For solid colours, click on the drop-down arrow in the **Colour** box to select from a limited number of colours or click on **More Colours** for a wider palette.
5 For gradients, choose a pre-set mix or select colours and then set the type and angle at which they merge together.
6 For pictures or texture, browse for a picture file or choose from the drop-down list of textures. You can also click on the **Clip Art** button to search for a Clip Art image.
7 If required, slide the transparency slider for a less opaque colour/picture.
8 To apply the background to a single slide, click on **Close**.
9 To apply the same background to the whole presentation, select **Apply to All**.
10 To return to the original background, click on **Reset**.

Click option

Browse for file

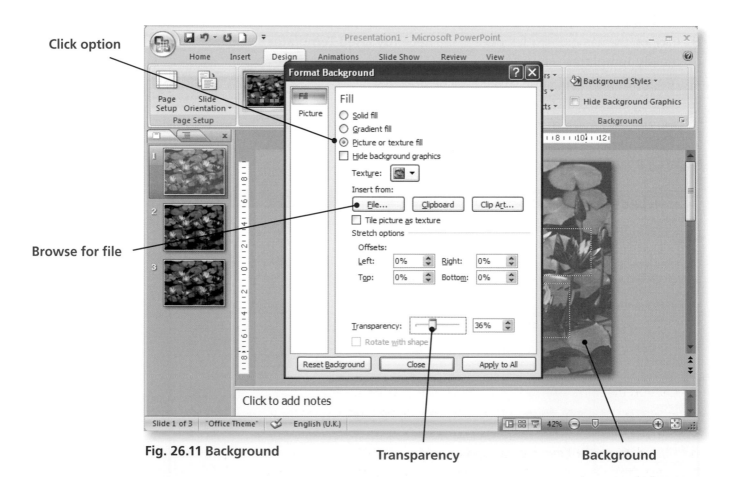

Fig. 26.11 Background   Transparency   Background

Check your understanding 3

1 Start a new presentation.

2 Save as *Clocks*.

3 Apply the image *clock picture* provided on the CD-ROM accompanying this book, as a background.

4 Set 50% transparency.

5 Save and close the file.

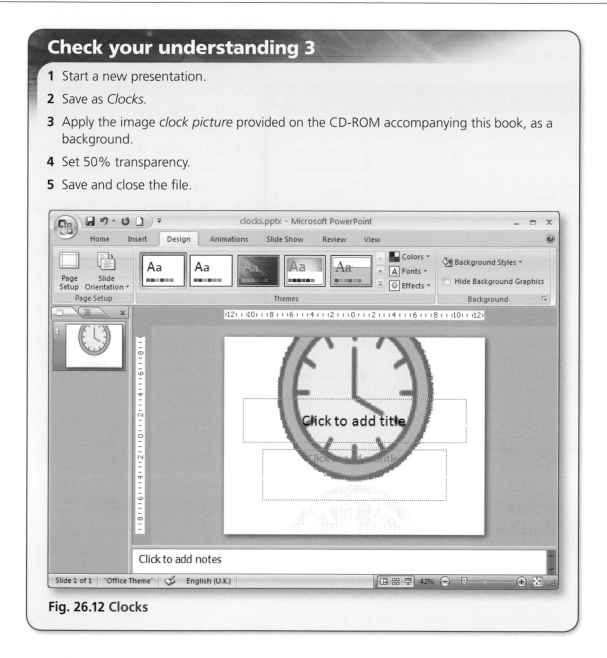

Fig. 26.12 Clocks

Headers and footers

If you want to add automatic entries such as the date or slide number, or type in your own text, you can do this by inserting footers in the bottom margin of the slides. You can also add header and footer entries to supporting documentation such as handouts or notes pages. The areas for different entries are pre-set on the slide, but you can drag the boxes to different positions and select and format entries using normal formatting tools.

add headers or footers

1 Click on **Header & Footer** on the **Insert** tab.

 Or

2 Click on the **Slide number** or **Date and time** commands.

3 When the Header & Footer box opens, click in the radio button for a fixed date and enter it in your preferred style.

4 For a date and/or time that will update automatically, click on the **update** option and select the style of entry you want from the drop-down arrow in the box.

5 Make sure the date is set as English (UK) or you will find the month precedes the day.

Check for time or date formats

Change to UK

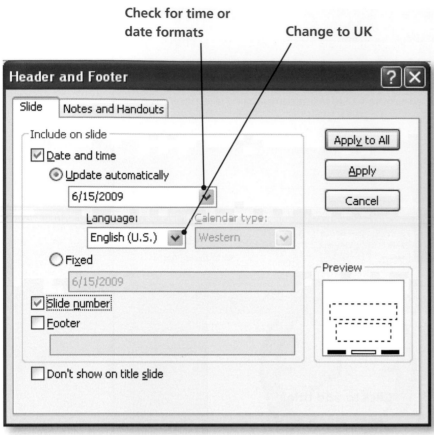

Fig. 26.13 Header footer box

6 Click on the **Slide number** checkbox to add numbers to your slides.

7 Click on the **checkbox** and then type in any other entries in the **Footer** box.

8 Click on **Apply** for a single slide.

9 Click on **Apply to All** if you want the same entries on every slide.

Check your understanding 4

1 Reopen the file *Clocks*.

2 Add the fixed date 12/2/2010 and a footer containing your name.

3 Select the entries and apply a bold emphasis.

4 Save these changes and close the file.

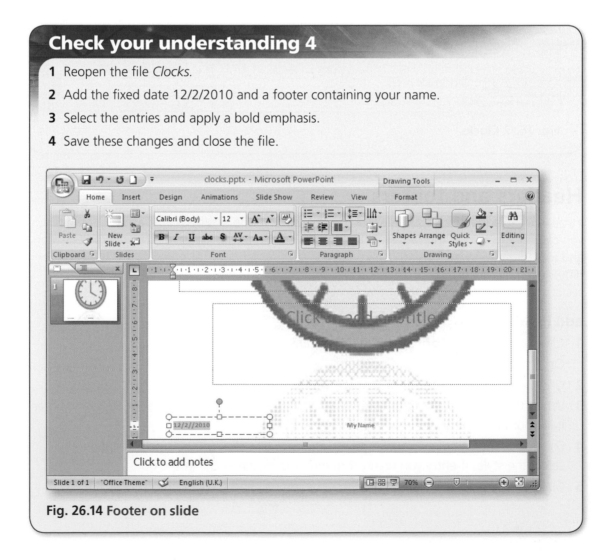

Fig. 26.14 Footer on slide

Text box properties

After adding text to a slide you can either display just the text or emphasise the text box/placeholder border or fill to make the text stand out more.

change text box properties

1 Click on the text to display the border and then right click on the line or inside the box.
2 The floating toolbar will appear and you can click on the drop-down arrow on the **Shape Outline** or **Shape Fill** button.
3 Choose a colour for the background or line and/or click on the arrow to open the gallery of line weights or styles.
4 Click on **More Lines** or **Colours** for further options.
5 Click on **No Outline or Colour** to remove the formatting.

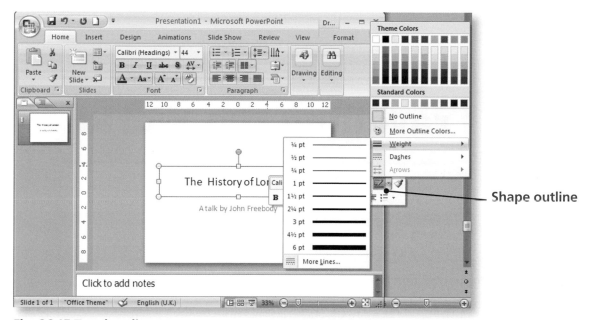

Fig. 26.15 Text box line

Or

6 Find the **Shape Outline** and **Fill** buttons on the **Format** tab.
7 If you right click and select **Format Shape**, you will open the **Format Shape** dialog box.
8 Click on **Line** or **Fill** and select colours and styles from the various boxes.

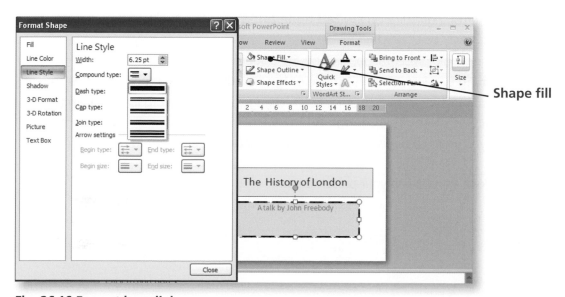

Fig. 26.16 Format box dialog

Check your understanding 5

1 Start a new presentation and save it as *Travel*.

2 Add the title *Travelling the World* to the first slide.

3 Apply a different font type and set the size to 50.

4 Shade the placeholder pale green.

5 Add a thick or double line border in red.

6 Save the file.

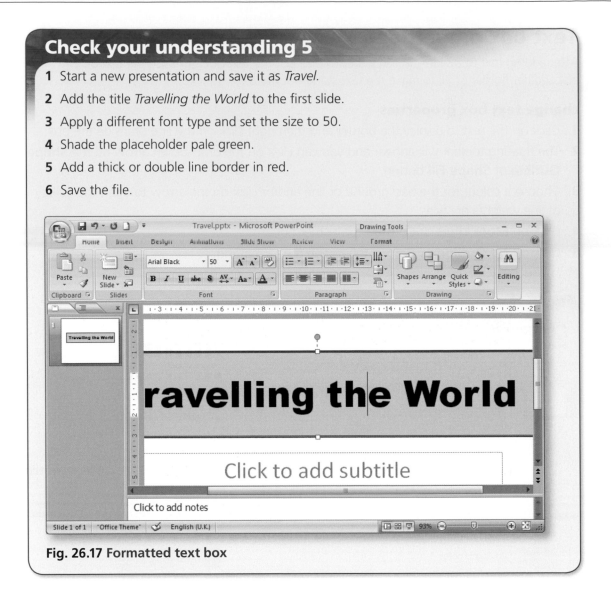

Fig. 26.17 Formatted text box

Spell checker

Once you have typed or copied any text onto a slide, you need to check that it is correct. Proofread by eye for words that are spelt correctly but don't make sense, and use the spell checker to change any misspelt words.

spell check a presentation

1 For a single error, right click on any **red underlined word** and select from alternative spellings that may be offered.

2 To check the whole presentation, click on the **Spelling** button on the **Review** tab.

3 In the **Spelling** window, click on **Ignore** or **Ignore All** to retain a spelling that is correct.

4 Click on a suggested alternative or manually change a misspelt word in the **Change to:** box and click on **Change** or **Change All** to update the slides.

5 For words you want recognised in future, click on **Add** to add them to the built-in dictionary.

6 Click on **Close** to return to your presentation.

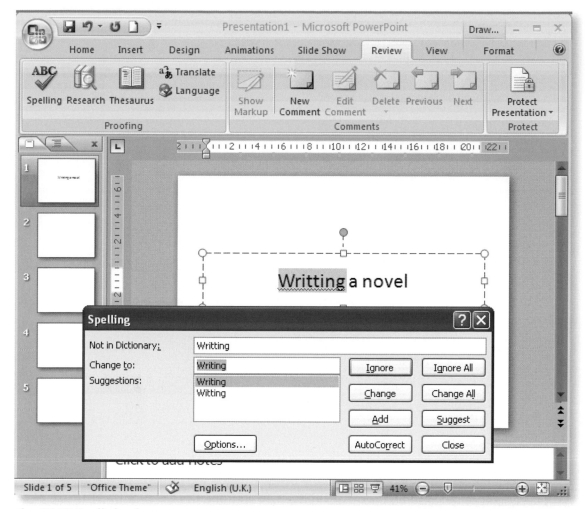

Fig. 26.18 Spell check

Bullet points

As you type into some placeholders, bullets will be added automatically. You can remove unwanted bullets or add them where needed. Each level of text has its own bullet style applied automatically. Numbering works in exactly the same way as bullets so you can number list items instead.

work with bullets

1 Select one line, or all list items.

2 Click on the **Bullets** button on the **Home** tab to add bullets.

3 Click on the highlighted **Bullets** button to remove bullets.

4 Click on the drop-down arrow next to the button to select alternative styles of bullet.

5 If you change text level within a list, each level will have a different style of bullet applied by default. It will also be indented.

- To go down a level, press the **Tab** key.
- To go up, hold **Shift** as you press **Tab**.

Bullets button

Numbering

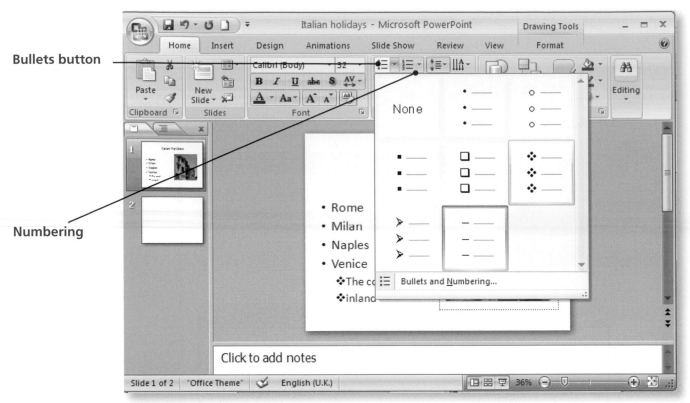

Fig. 26.19 Add bullets

6 Sometimes the bullet points or list items will be in the wrong place on the slide. For example, the text may be too close to the bullet. Change the position by dragging the bullet point or text insertion point along the ruler with the mouse. A dotted line will show the new position.

7 To increase spaces between entries in a list, apply different line spacing.

Line spacing

Bullet point position

New list item position

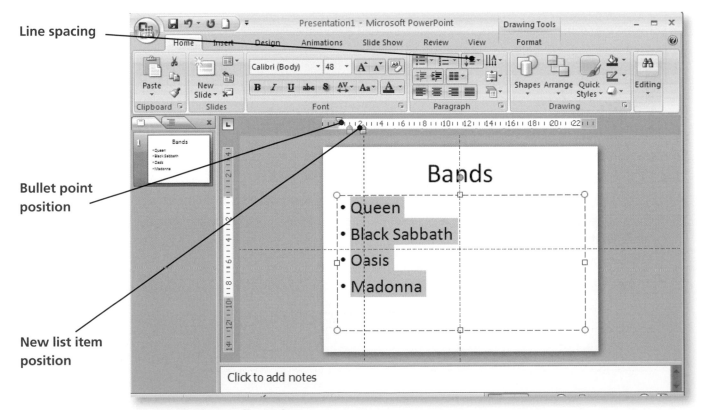

Fig. 26.20 Change list indent

add extra items to a bulleted list

1 Click at the end of the item above and press **Enter**.

2 Type in the new entry.

3 To create an entry at a lower level, first press **Enter** to move to the line and then press the **Tab** key.

4 Select individual level entries to apply a different bullet style.

Check your understanding 6

1 Reopen *Travel*.

2 Add the following bulleted list to the slide, left aligned, in the subtitle placeholder:

- UK
- Europe
- Africa
- India
- The Far East

3 If necessary, drag the title placeholder higher up the slide to make room.

4 Change the style of bullets that have appeared by default.

5 Apply a Courier font size 36 to the list text.

6 Now add two items with a different style of bullet, indented at a lower level under Africa:

- Egypt
- South Africa

7 Save the file as *Travel bullets*.

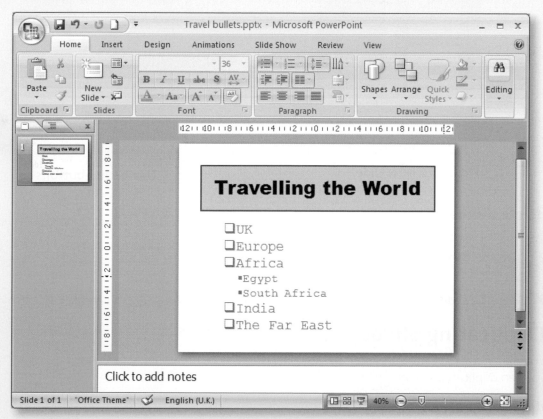

Fig. 26.21 Travel bullets

Building up a presentation

Most presentations are made up of a number of slides and these can be reordered, duplicated or deleted as you work on their contents.

add new slides

1 In the **Slide** pane, click on a thumbnail and press **Enter**.

Or

2 Click on the **New Slide** button.

3 Choose a layout and the new slide will appear as the one *after* the slide on screen.

4 Move between slides by clicking on the number in the **Slide** pane or using the **Previous** and **Next** navigation buttons on the right-hand side of the slide.

Add slide

Delete slide

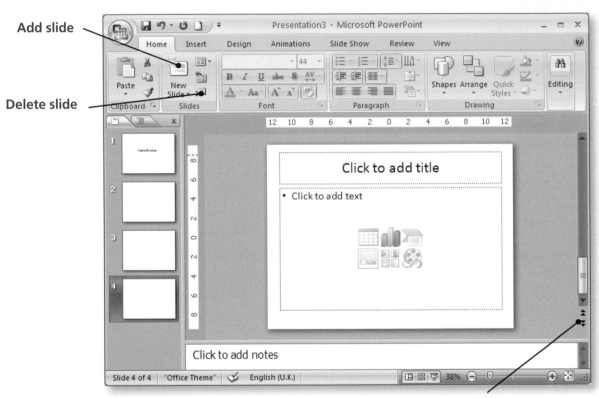

Fig. 26.22 New slide

Navigation buttons

delete a slide

1 Click on the slide in the **Slide** pane.

2 Press the **Delete** key.

Or

3 Open the slide on screen and then click on the **Delete Slide** button showing a red cross.

Duplicating slides

You may want to create slides that are very similar. Rather than designing each one from scratch, you can duplicate them and then make the minor changes necessary.

duplicate a slide

1 Select the slide(s) you want to duplicate in the **Slide** pane.

2 On the **Home** tab, click on the drop-down arrow below the **New Slide** command.

3 Select **Duplicate Selected Slides**.

4 Copies of the slide(s) will appear as the next slide(s) in the presentation.

5 To copy a slide from one presentation to another, use **Copy and Paste**.

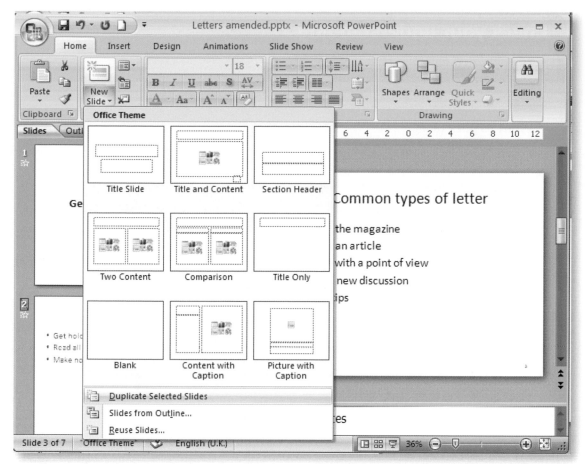

Fig. 26.23 Duplicate

Reordering slides

Having added a number of slides and decided on their content, you may want to display them in a different order. You can change slide order in two views: on the **Slide** pane in **Normal view** or in **Slide Sorter view**.

New position for slide

Fig. 26.24 Reorder slides

Slide sorter view

reorder slides

1 Click on the slide you want to move on the **Slide** pane or in **Slide Sorter view**.

2 Hold down the mouse button.

3 Drag the slide to its new position.

4 This will be marked by a red vertical or black horizontal line.

5 Let go of the mouse and the slide will drop into place.

Check your understanding 7

1 Reopen *Travel bullets.*

2 Add two new slides.

3 Add the following text to slide 2:

What to pack (title)

- Clothes
- Documentation
- Washing things
- Things to do

4 Add the following text to slide 3:

Where to go (title)

- Museums and galleries
- Train or boat trips
- The sights
- Towns and villages

5 Now reorder the slides so that slide 3 becomes slide 2.

6 Duplicate slide 1 and move this slide to become the last slide in the presentation.

7 Now delete the duplicated slide.

8 Save the file as *Travelling* and close.

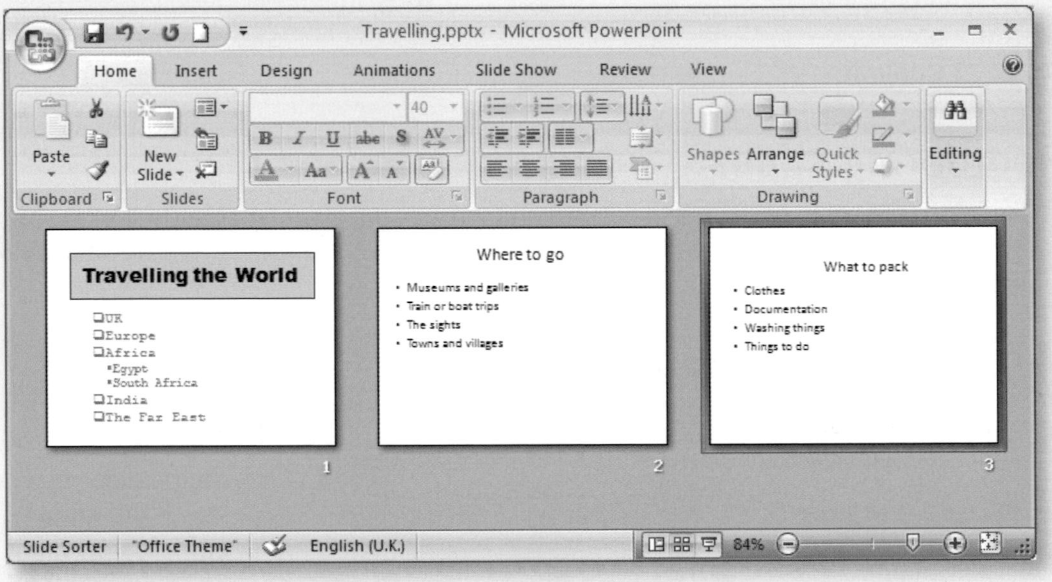

Fig. 26.25 Travelling

Find and replace

As with all Microsoft Office programs, there are useful search tools available to help you locate or make changes to text in your presentation. In PowerPoint they work in exactly the same way as in Word but there are less options to choose from.

find data

1 Click on the **Find** command on the **Home** tab in the **Editing** section.

2 Enter the text you want to find in the **Find what:** box.

3 You can click in the checkboxes to match case exactly and/or to find whole rather than parts of entries.

4 Click on **Find Next** to locate the first matching entry and keep clicking on the button to work through the presentation.

replace entries

1 Click on the **Replace** command on the ribbon or click on the button in the **Find** dialog box.

2 Type in the entry to be replaced in the **Find what:** box.

3 Enter the replacement text in the **Replace with:** box.

4 Set any options such as matching case.

5 Click on **Replace All** if you are sure you have entered items correctly.

6 Click on **Find next** to locate the first matching entry.

7 Click on **Replace** to replace it or **Find Next** to leave it in the presentation and move on to the next matching entry.

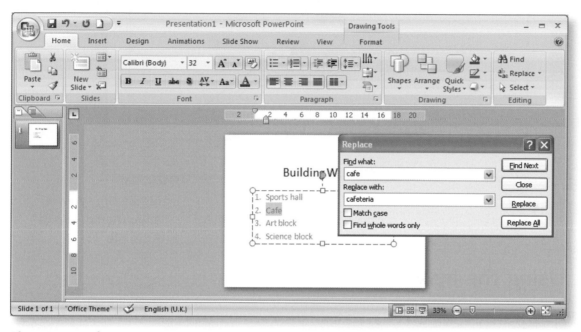

Fig. 26.26 Replace

Adding objects to slides

There are a number of additions you can make to your presentation to bring it to life or make the information it contains clearer or more memorable. These include:

- graphical images
- WordArt (which enables text to be shaped and textured)
- drawn shapes
- charts
- tables
- diagrams
- sounds
- animated images (labelled 'movies' in Microsoft programs).

Note that sounds and animations will only work in slide show view.

Depending on the type of object, once it is on a slide it may be possible to resize, rotate, align, layer, copy, move or group it with other objects, or format it by adding coloured borders or fills.

Content placeholders

When you apply a slide layout containing a **Content** placeholder, you will see small icons representing many of these objects. Click on any of the icons to start inserting that particular type of object. You will be able to either create the object directly or search the computer for files of that type.

Create a table

Create a chart

Create a SmartArt graphic e.g. organisation chart

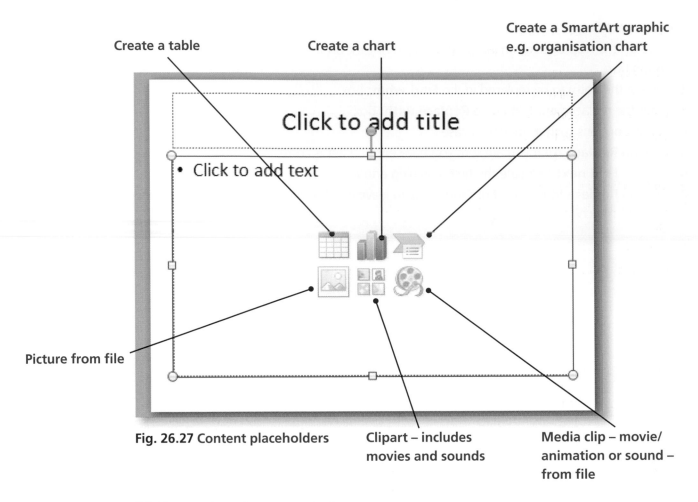

Picture from file

Fig. 26.27 Content placeholders

Clipart – includes movies and sounds

Media clip – movie/ animation or sound – from file

Using the Insert tab

If you click on the **Insert** tab you will see many of the objects listed. Click on the command to open up an appropriate dialog box or menu.

Object to insert

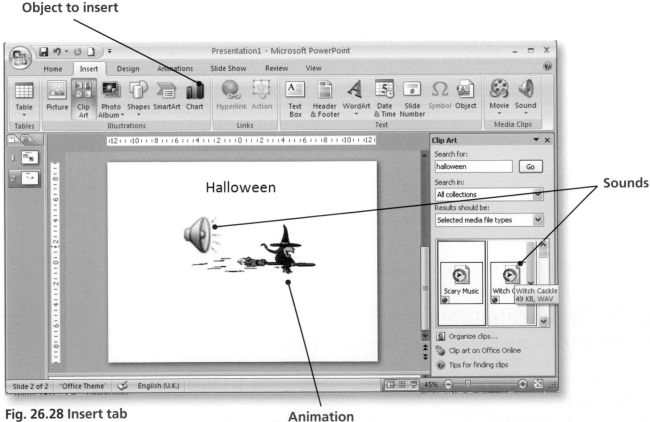

Sounds

Animation

Fig. 26.28 Insert tab

insert a Clip Art image

1 Click on the **Clip Art** command on the **Insert** tab.
2 Click on the **icon** in a **Content** placeholder.
3 A search pane will appear.
4 Enter your keywords and click on **Go** to locate appropriate images.
5 Scroll through the images.
6 Click on an image to add it to your slide.

insert a picture from file

1 Click on the **Insert Picture** from file command.
  Or
2 Click on the **Content placeholder icon**.
3 Browse through your folders to locate the file.
4 Click on its name and then click on **Insert**.

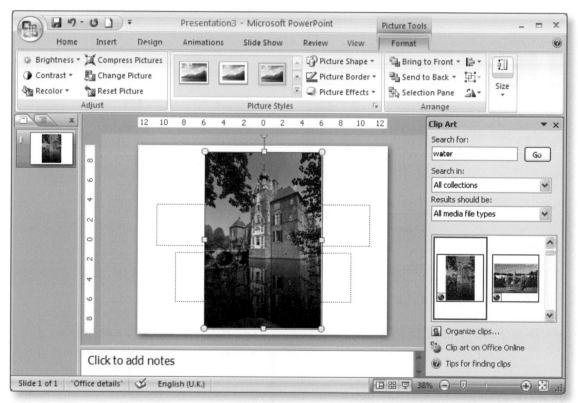

Fig. 26.29 Insert picture

Copy and Paste

You will often find it quicker to copy and paste in an image. For example, any copyright-free images from the Web can be saved onto your computer or storage media to be inserted from file, or you can copy and paste them in directly.

Formatting pictures

When a picture appears on the slide, it is in a selected state and shows small, white sizing handles round the edge plus a green rotate circle on an arm.

resize a picture

1 Make sure it is selected.
2 Position the pointer over a sizing handle. Select a corner position to keep the picture in proportion.
3 Gently click and hold down the mouse button as you drag the border in or out.

4 Let go when the image is the correct size.

Or

5 On the **Picture Tools – Format** tab, enter exact measurements in the height or width boxes in the Size group.

Or

6 Right click on the picture and select **Size and Position** to open the dialog box.

7 Click on the **Size** tab and change height or width measurements.

8 You should see that there is a tick in the **Lock aspect ratio** box which will maintain the picture's proportions.

Sizing handle   **Set measurements**

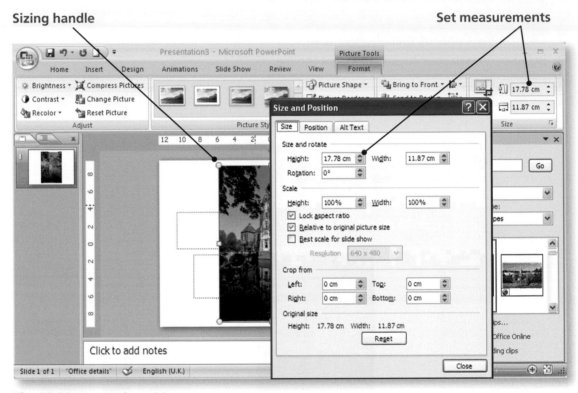

Fig. 26.30 Size and position

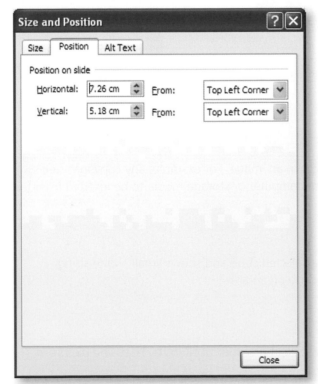

Fig. 26.31 Position

position a picture exactly

1 Open the **Size and Position** dialog box.

2 Click on the **Position** tab.

3 The object can be positioned with a set measurement from the top left-hand corner (or centre) of the slide horizontally and/or vertically.

move a picture

1 Gently drag the picture to a different position when the pointer shows a four-headed arrow.

Or

2 Right click and select **Cut**.

3 Click on the same slide or select a different slide to receive the picture.

4 Right click and select **Paste**.

copy a picture

1 Right click and select **Copy**.

2 Click on the slide to take off the selection, or click on a new slide.

3 Right click and select **Paste**.

4 You can also drag the picture as you hold down the **Ctrl** key to create a copy on the same slide.

delete a picture

1 Select the picture and press the **Delete** key.

Cropping

If a picture includes an unwanted portion, this can be removed by a method known as **cropping**. (It is not the same as resizing as that leaves all parts of the picture still visible.)

crop a picture

1 Select the picture.

2 Click on **Crop** on the **Format** tab.

3 This places black lines round the picture and the pointer will show the cropping shape.

4 Move the pointer to the border nearest the unwanted part and click on and drag the black shape.

5 Slowly drag this inwards or upwards.

6 Let go when the unwanted part of the picture disappears.

7 You can now work with what is left.

8 If you make a mistake, click on the **Crop** command again and drag the boundary back to its original position.

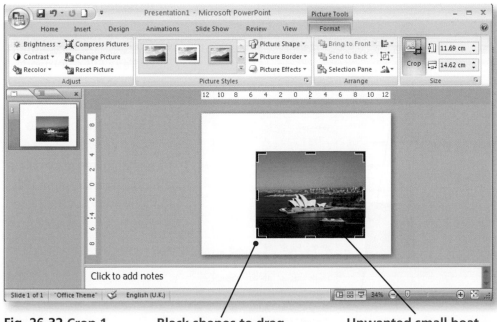

Fig. 26.32 Crop 1   **Black shapes to drag**   **Unwanted small boat**

Small boat removed

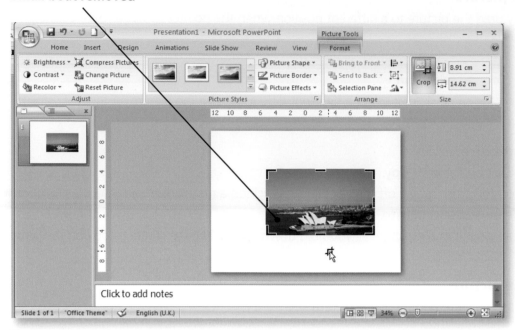

Fig. 26.33 Crop2

Adding WordArt

When you want your text to stand out, you can create a WordArt object rather than rely on formatted normal text. This object can then be moved, copied, deleted, positioned or formatted in a similar way to graphical images.

insert WordArt

1   Click on the **WordArt** command on the **Insert** tab.
2   Work through the following steps:

   a. Choose a style of font from the gallery.

   b. Type in your text.

   c. On the ribbon, click on the **Text Effects** command and click on **Transform** to change the shape of the object or apply a shadow or other effect.

Text fill

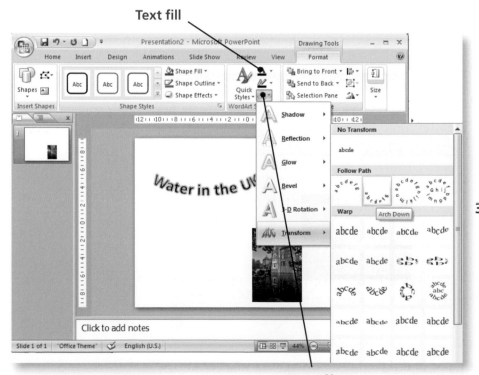

   d. Click on **Quick Styles** to reopen the gallery of fonts.

   e. Click on the **Text Line** or **Text Fill** commands to set different colours for the words.

   f. Click on the object to edit the text.

3   Move, copy, resize or delete the object in the same way that you work with pictures.

Fig. 26.34 Shape WordArt   **Text Effects**

Check your understanding 8

1. Start a new presentation and save it as *Sailing*.
2. Add a WordArt title that displays the text *Boating for Life.*
3. Apply a curving shape and select an appropriate font or fill the shape so that the text is a blue colour.
4. Move it to the top of the slide.
5. Change the text to read: *Sailing for Life.*
6. Now insert any Clip Art picture to do with boats or sailing.
7. Make the image 7cm in height.
8. Position it in the centre of the slide.
9. Copy the image and place the copy in the top left-hand corner. Make sure it does not obscure the WordArt.
10. Resize the copy so that it is now only 3.5 cm in height.
11. Save and close the file.

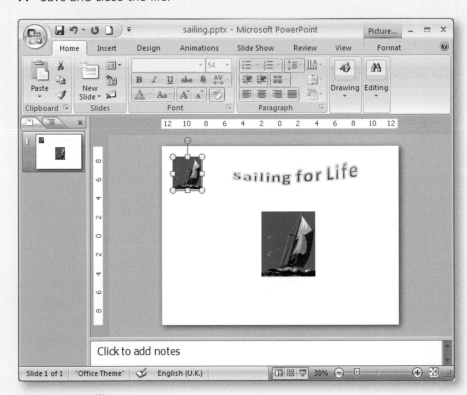

Fig. 26.35 Sailing

Drawn shapes

PowerPoint 2007 contains a range of shapes and lines that you can add to a slide very quickly. You can then resize, rotate, format or group the shapes to build up complex and attractive drawings.

add a line or shape

1. On the **Home** tab in the **Drawing** group, or from the **Insert** tab, click on the **Shapes** command.
2. Click on any style of line or shape to select it.
3. Click on the slide to add a shape or line at the default size.
4. Move the pointer to the slide and click and drag out the shape to set your own size.
5. When you let go, the shape will appear selected with a blue fill by default.

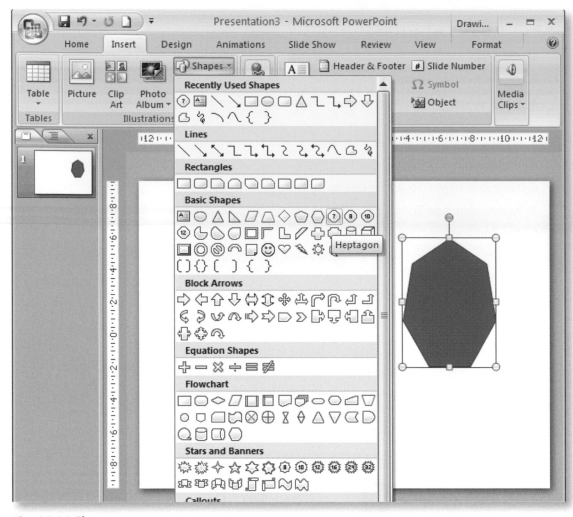

Fig. 26.36 Shape

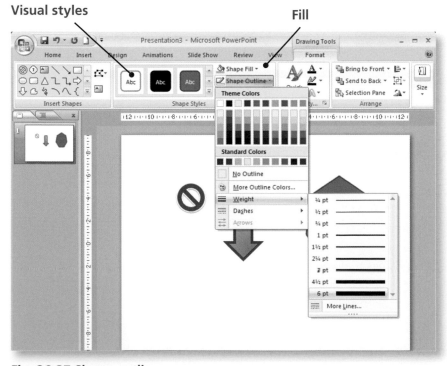

Fig. 26.37 Shape outline

format a shape

1 Select the shape.

2 Click on the **Format** tab or use commands on the **Home** tab in the **Drawing** group.

3 Rest the pointer on one of the visual styles in the gallery to preview the effect and click on any to set a pre-defined design to your shape.

4 If you prefer, choose a different colour fill or line style and weight from the drop-down **Shape Fill** and **Outline** boxes.

5 For an arrow (not a block arrow), click on **Arrows – More Arrows** from the **Shape Outline** box to change the beginning or end type or size of arrow head.

6 Click on **Effects** to apply shadows or reflections.

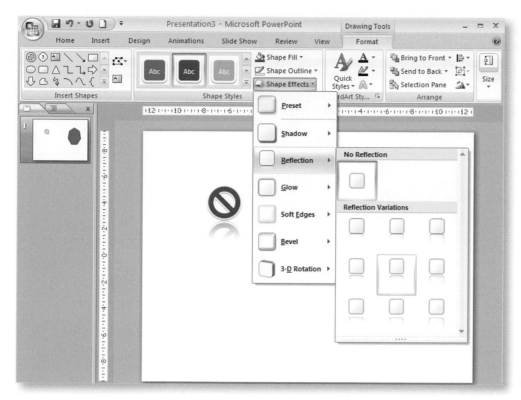

Fig. 26.38 Format shape

Adding text to shapes

If you want a message inside a shape, it is easy to add text.

add text to a shape

1 Click on the shape and enter your text.
2 On the **Home** tab, click on the **Text direction** box to select **vertical** or **stacked** text.
3 Click on **More Options** to set the text direction more exactly.
4 To edit the text, click on it and delete or amend as normal.
5 You can also select it and change alignment, font, font size or colour from commands on the **Home** tab. (A **Text Alignment** button is below **Text Direction**.)

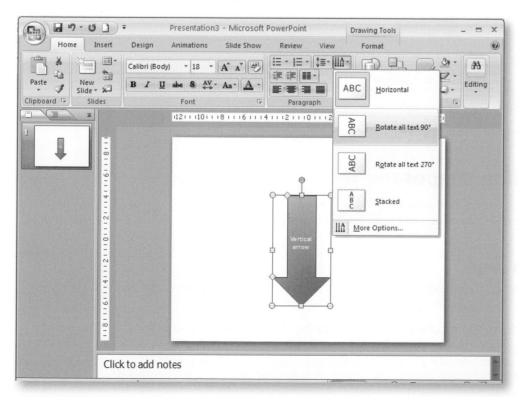

Fig. 26.39 Text in shape

Check your understanding 9

1 Start a new presentation.

2 Add a circle, square and triangle of any size.

3 Make the circle quite small and colour it red.

4 Colour the square yellow.

5 Make the triangle very large, colour it green and border it with a thick, black solid line border.

6 Join the shapes with single arrows.

7 Increase the line weight for the arrows and colour them pink.

8 Add the text *Simple shapes* inside the triangle and position it so that it reads vertically upwards.

9 Increase the text font to size 36.

10 Close the file without saving.

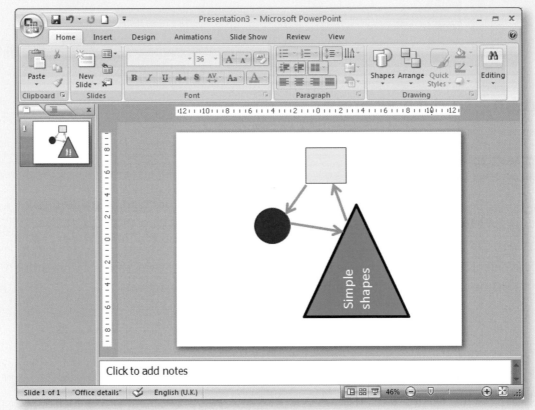

Fig. 26.40 Simple shapes

Rotating shapes

A selected shape shows a rotate arm that ends with a green circle. You can use this to drag the shape round manually, or set an angle of rotation. You can also flip shapes to create mirror images.

rotate a shape

1 Select the shape and position the pointer over the green rotate circle. You will see a black circle.

2 Hold down the mouse button and drag the shape round to a new position. The pointer will now display four black arrows.

Or

3 Click on the **Rotate** command on the **Format** tab.

4 Select an option such as **Flip Vertical** to invert it, **Flip Horizontal** for a mirror image or rotate the shape in one direction by 90 degrees.

5 Click on **More Options** to open the dialog box where you can enter an exact angle of rotation.

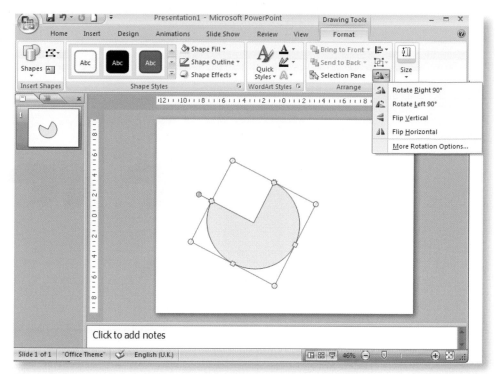

Fig. 26.41 Rotate

Multiple shapes

Once you have added a number of shapes to a slide, you can work with them in several ways:

- Align them to each other or the edges of the slide.
- Group them into a single drawing.
- Layer them on top of one another.

align shapes

1 Select one or more shapes.

2 Click on the **Align** command on the **Format** tab.

3 Select an alignment option.

4 For example, **Align Top** will move the selected shapes up to the top of the slide.

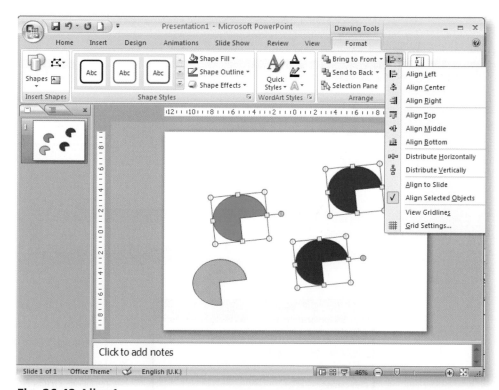

Fig. 26.42 Align1

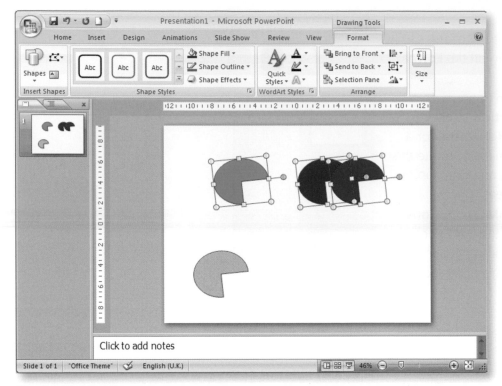

Fig. 26.43 Align2

Drawing guide   Grid

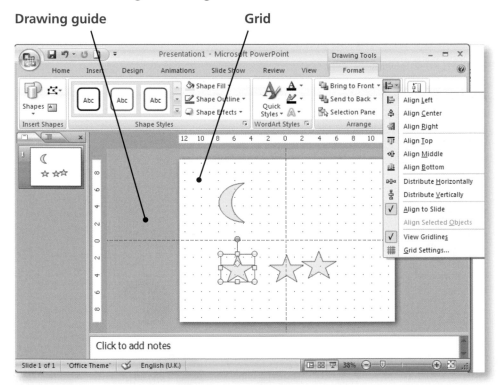

Fig. 26.44 Use grid

5 To help position objects by eye, you can display four drawing guides across the slide, or a set of gridlines. You can then drag objects and line them up with the lines and rulers.

6 If no rulers are visible, click on the checkbox on the **View** tab.

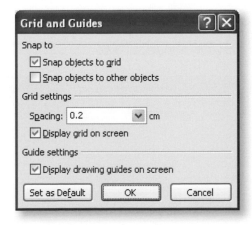

Fig. 26.45 Grid and guides

7 To open the **Grid and Guides** dialog box, right click on the slide or click on **Grid settings** from the **Alignment** command.

8 Take off the tick in the boxes if you want to remove grids and guidelines.

group shapes

1 Select the shapes you want to group. Either click on the first and then hold down **Ctrl** as you click on the others or draw round them with the pointer.

2 You can also include other objects such as WordArt or text boxes and group them with shapes.

3 Click on the **Group** option from the **Group** command on the **Format** tab.

4 You will see that all the shapes have been grouped into a single selected box and can be treated as one to copy, move or resize.

5 You will still be able to click on and select an individual shape to reformat it, even after it has been grouped.

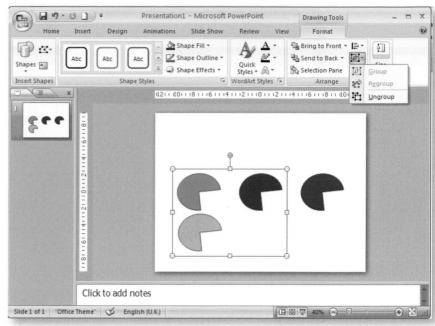

Fig. 26.46 Group shapes

6 As an alternative, you can click on **Ungroup** to make changes to parts of the grouped drawing, and regroup them again by clicking on **Regroup**.

Layering

An alternative way to build up a complex drawing is to layer shapes on top of one another. Once they are in place, an individual shape can be repositioned – either sent behind a single shape (backwards) or to the back of the stack, or brought forward in the same way. If you set a transparency level, you will be able to see shapes below others that have been layered. This is a useful way to use coloured shapes to frame text or other objects.

layer shapes

1 Drag the shapes on top of one another.

2 Make sure you can select a small part of any shape.

3 Select one shape and reposition it by sending it behind or in front of one or all the other shapes. Either click on **Bring to Front** or **Send to Back** and select **to the back/front** or **backwards/forwards**.

Or

4 Click on the **Selection Pane** command, click on a shape to select it and then move it backwards or forwards by clicking on the up or down arrows.

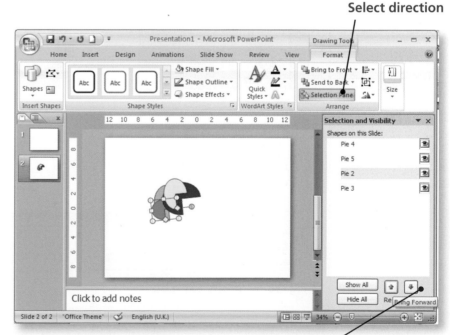

Fig. 26.47 Reorder shapes

Click to move selected shape

5 To set transparency, select a shape and then click on the **Shape Fill** command and select **More Fill Colours**.

6 Set the transparency slider manually or enter a percentage – 100% will make it completely transparent.

Check your understanding 10

1 Start a new presentation that has two slides with blank slide layouts.

2 On slide 1, insert four circles and colour them black, white, light grey and dark grey.

3 Position them centrally in the four quarters of the slide by eye using the gridlines, guides and rulers to help you.

4 Now use the alignment options to position them exactly.

5 Group the shapes into one.

6 Copy the grouped shape to a second slide.

7 Colour the dark grey shape red.

8 Ungroup and then layer the shapes in the following order: black on top, then red, then grey, then white. Make sure part of each shape is available for selection.

9 Reorder the white shape so that it is one from the bottom.

10 Make the black shape 50% transparent.

11 Add an irregular shape such as a moon of any size to a separate part of the slide and colour it green.

12 Copy this shape and flip it to create a mirror image.

13 Position the two moons opposite one another aligned at the top of the slide.

14 Save the presentation as *Shape exercises*.

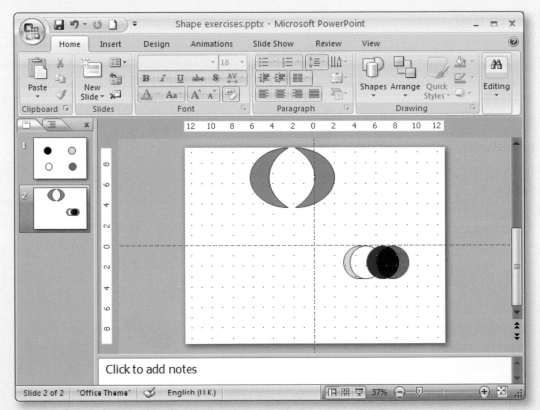

Fig. 26.48 Shape exercises

Tables

To add a table to a slide, you use the same process as when creating a table in a word processed document. Once it appears, it can be resized, formatted and edited.

When the table appears, it will have a format set by default. On the **Table Tools – Design** tab in the **Table Styles** section you will find a range of different designs in the gallery that you can apply. You can also choose **No Style, Table Grid** for a plain table.

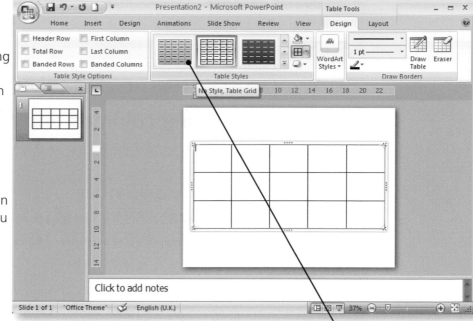

Fig. 26.49 Table style

Table styles

insert a table

1 Click on the **Table** command on the **Insert** tab.
2 Drag the mouse across the cells to select the correct number of rows and columns.
3 Let go of the mouse and the table will appear.
   Or
4 Click on the **Table** icon in the **Content** placeholder and enter the number of rows and columns into the boxes.

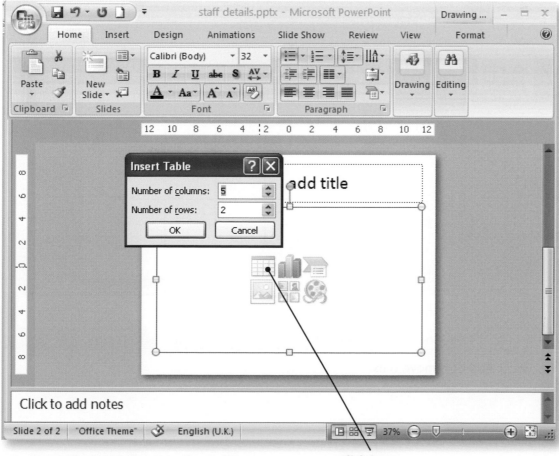

Fig. 26.50 Insert table using placeholder

Click icon

make overall changes

1 Move the table to a new position on the slide by dragging it with the mouse.

2 Resize the table by dragging out a border.

3 There is also a button on the **Table Tools – Layout** tab where you can resize the table exactly by typing in a new height and width.

Adding or removing columns and rows

If you need to make changes to the table you can do so in a number of ways.

add columns or rows

1 Click on a cell.

2 On the **Table Tools – Layout** tab, click on an **Insert** option such as **Above** or **Right**.

3 You can also click on in the last cell and press the **Tab** key to add a new row.

remove columns or rows

1 Drag to select the unwanted cells.

2 Click on the correct option on the **Table Tools – Layout** tab after clicking on the **Delete** command.

3 Select the table border and press the **Delete** key on the keyboard to remove the table.

Click for options

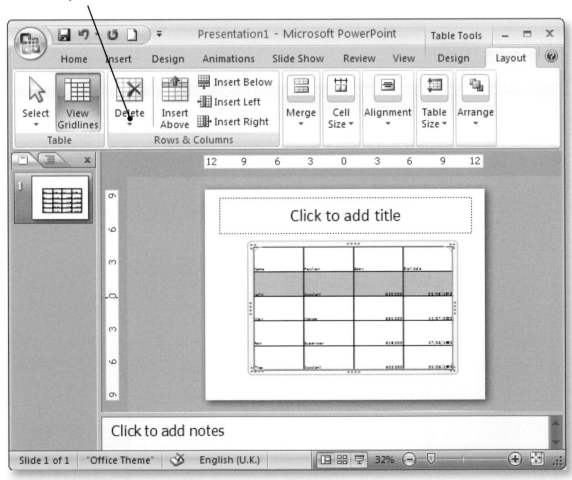

Fig. 26.51 Add or remove cells

Adding text

There are two ways to add text into a table:
- Type into each cell in turn.
- Copy across data from elsewhere.

copy in data from a data file

1 First create a table of the required size on a slide.
2 Open the data file containing the data and select all relevant cells.
3 Click on **Copy**.
4 Return to your slide and either click on a single cell or select all the table cells by dragging across with the mouse. They will appear light blue.
5 Click on **Paste**.

Formatting a table

As well as changing the appearance of data entered into the cells, you can also change the appearance of a table by adding different borders and shading, or removing borders altogether.

format cells

1 Select the cells containing the data.
2 Use normal formatting tools on the **Home** tab to apply different font types, sizes or to colour the text.
3 Use the alignment buttons to place the text horizontally in the cells. (See Unit 022 for details of the method for using tabs in tables to align data within cells.)
4 For a deep cell, you can click on the **Align Text** button to set text vertically in the middle, top or bottom.
5 Drag cell boundaries to increase or decrease measurements, or enter exact measures into the cell size boxes on the **Layout** tab.

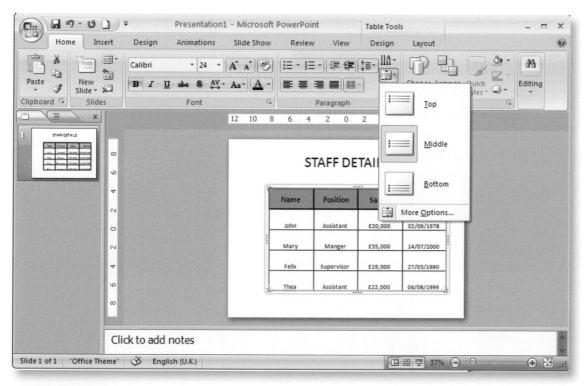

Fig. 26.52 Align text in cell

6 On the **Table Tools – Design** tab, set a type of border from the border button.

7 To remove borders, choose the **No Border** option.

8 Use the line style and width buttons in the **Draw Borders** group to change the type of line, making sure you click on the border button to apply the selection.

9 Add background colours to the cells from the **Shading** button.

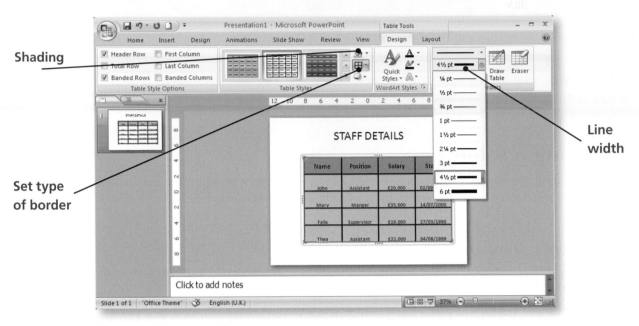

Shading

Set type of border

Line width

Fig. 26.53 Border table

Check your understanding 11

1 Start a new presentation.

2 On slide 1, create a table that has four columns and five rows.

3 Increase the table size so that it fills the slide.

4 Copy across the data from the Excel file *Staff List* provided on the CD-ROM accompanying this book.

5 Add the slide title *Recent staff changes.*

6 Format the column headings to Arial font size 24.

7 Reduce the depth of these cells to about 1.5 cm and centre align the entries.

8 Format all other data to Arial font size 18.

9 Format the borders to a solid line with a width of 2 ¼ pt.

10 Shade only the cells containing the names green. All other cells should have no shading.

11 Save the file as *Staff details.*

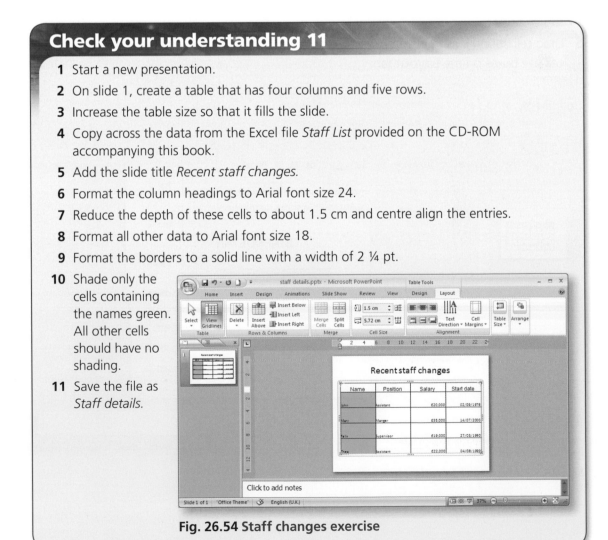

Fig. 26.54 Staff changes exercise

Charts

You will often want to display data on a slide in the form of a chart, and you can do this either by creating the chart directly or by copying across a chart from an application such as Excel.

create a chart on a slide

1 Apply a content slide layout and click on the **Chart** icon.
   OR
2 Click on the **Insert** tab and select **Chart**.
3 Select the appropriate chart type in the gallery that appears and click on **OK**.
4 You will be presented with a ready-made chart sample together with a spreadsheet containing temporary data.
5 Replace the data in the spreadsheet with your own and the chart will be created for you.
6 If you are offered too many columns or rows in the spreadsheet, select and delete them or they will confuse your final chart.
7 To amend the data at any time, reopen the spreadsheet by clicking on **Edit Data** on the **Design** tab or after right clicking.

Reopen spreadsheet

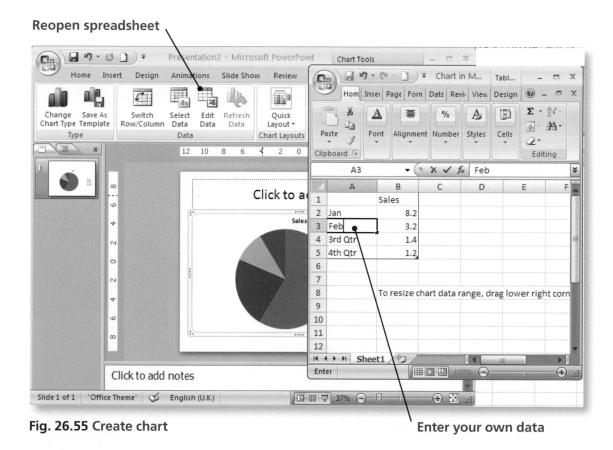

Fig. 26.55 Create chart

Enter your own data

Formatting the chart

The chart is created using Excel and so you can use the familiar methods for making changes, such as:

- adding labels
- adding or removing the legend
- adding a chart title
- formatting colours of the data series
- changing chart type.

You can resize the chart on the slide by dragging out a border, or move it by dragging it with the mouse.

Add labels

Change chart

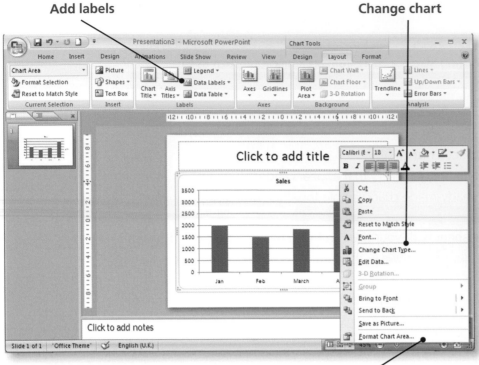

Fig. 26.56 Format chart

Format selection

format the chart

1 Click on the **Chart Tools – Layout** tab and add data labels or titles by clicking on the relevant buttons.

2 Right click on the chart or click on the command on the **Design** tab to select a different chart type.

3 Right click on a selected part of the chart or click on the **Format** tab to change colours and line styles.

Check your understanding 12

1 Start a new presentation and save the file as *Best sellers*.

2 Create a bar chart using the following data:

Event	Tickets
Christmas fair	350
Halloween party	56
Summer fete	426
Dog show	245

3 Remove the legend.

4 Change the title to *Ticket sales for best selling events*.

5 Increase the font size and make sure the title is well above the chart data.

6 Add X and Y axes titles: *Event* and *Ticket sales*.

7 Colour the data series red and the plot area yellow.

8 Save and close the file.

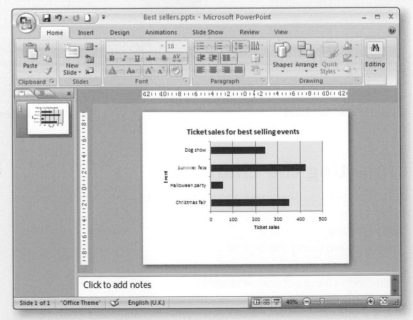

Fig. 26.57 Best sellers

Organisation charts/diagrams

A different type of chart you may want to add to a presentation is one that represents a hierarchy or family tree. Each person or item is displayed in its own box and these can have a particular relationship to other boxes in the chart.

insert an organisation chart

1 Click on the **Insert SmartArt** graphic placeholder.

    Or

2 Click on this option on the **Insert** tab.

3 Choose **Hierarchy** in the index and select your preferred layout. The first option is **Organisation Chart** and the second is **Hierarchy**.

4 When the chart appears, click into each box and enter your own text. The text will be resized to fit the box automatically.

5 Press **Enter** for entries on a new line in the same box.

Organisation chart

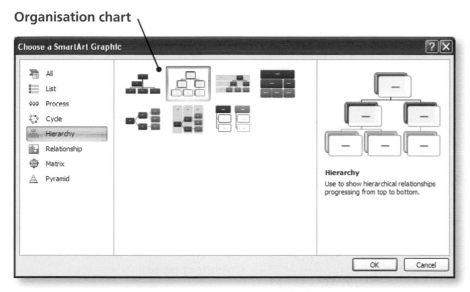

Fig. 26.58 Insert hierarchy

6 To add new relationships, click on a box and then click on the **Insert Shape** button on the **SmartArt Tools – Design** tab.

7 For the Organisation Chart, click on the drop-down arrow below the **insert Shape** command to add people in roles such as Assistant (below but not directly below), Co-worker (at the same level) or Subordinate (directly below).

8 Promote or demote a new box if you want to move it to a new level by clicking on the appropriate arrow button in the **Create Graphic** group.

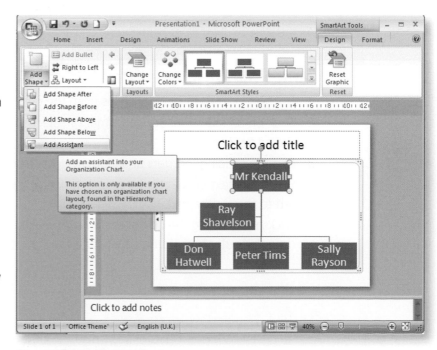

Fig. 26.59 Add assistant

9 You can also click on the **Text Pane** button to type in box entries directly.

10 Click on any box and press the **Delete** key to remove it from the chart.

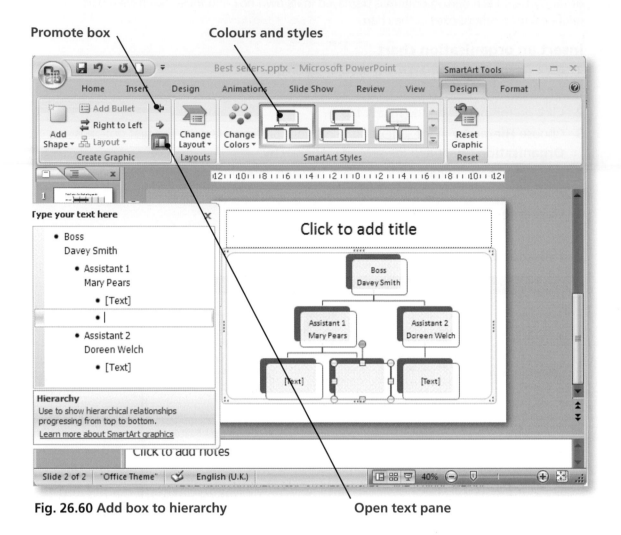

Fig. 26.60 Add box to hierarchy   **Open text pane**

Formatting the organisation chart

As the style of chart is set by default, you can select an alternative from the gallery or change individual boxes manually.

format the organisation chart

1 Set new colours or general styles from the gallery on the **Design** tab.

2 Click on an individual box to format or select several boxes by drawing round them with the arrow pointer.

3 Change shading, borders or shadow effects from the **Format** tab.

4 Select the text in any box (or several boxes at the same time) and use tools on the **Home** tab to format the text.

5 Click on and drag the box boundaries to make them larger or smaller, and click on and drag the outer blue border to change the size of the entire chart.

Check your understanding 13

1 Start a new presentation and save it as *Department*.

2 Create a slide showing the following organisation chart:

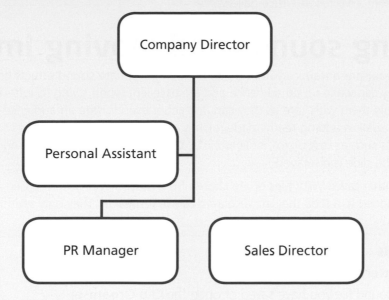

3 Now add a new Manager at the same level as *PR Manager* and *Sales Director* and add these details: *Head of Finance*.

4 Add a new linked box at an Assistant level: *Assistant to Head of Finance*.

5 Give the slide the title *Company Structure*.

6 Finally, remove the *PR Manager* role from the chart.

7 Save and close the file.

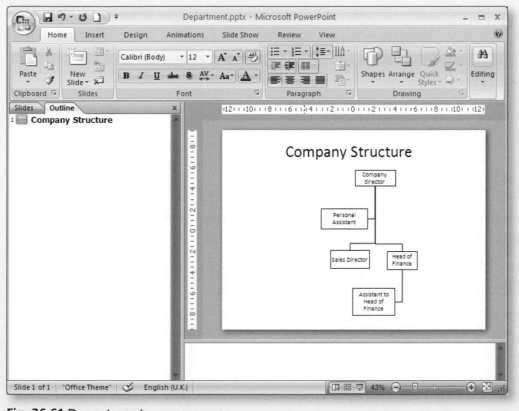

Fig. 26.61 Department

Creating an organisation chart using basic shapes

Instead of using the SmartArt graphics, you could draw your own chart by inserting shapes and lines from the main shape gallery. Clearly this is very labour-intensive as each box and line must be sized and placed correctly, and so it is not recommended unless the style of chart is definitely unavailable from a Microsoft Office gallery.

Adding sounds and moving images

To enhance your presentation, you may like to add appropriate sound effects or animated pictures. They can wake up an audience and are excellent when trying to entertain, but you need to handle them with care as they can distract or even irritate an audience if overused. (Note that as well as adding ready-made movies, the term **animation** applies to presentations where objects such as text boxes, bulleted lists, titles or charts are animated and build up piece by piece as the slide is displayed.)

Sounds are either saved WAV files or are chosen from those stored with the Clip Art or Media files. When added to a slide they are displayed as a loudspeaker. They only sound when you run a slide show.

add sounds

1 On the **Insert** tab, click on **Sounds**

2 Select a sound file you have saved or open the **Clip Organiser**.

3 If you use the placeholder in a **Content slide** layout, it will take you to your computer files.

4 You can also go to **Insert – Clip Art** and make sure you select **Sounds** in the **Results should be:** box.

5 Enter keywords or scroll through the available sounds. These include bells, telephones, clapping and songs.

6 Click on any sound to add it to the slide. it will appear as a loudspeaker symbol.

7 Choose whether to hear the sound automatically or only when you click on the icon on the slide.

Fig. 26.62 How to hear sound

8 Click on the **Slideshow** button to hear the sound.

9 If you have chosen the automatic option, click on **Escape** to cancel the sound and close the slideshow or **Page Down** to move to the next slide.

Loudspeaker symbol

Select sounds

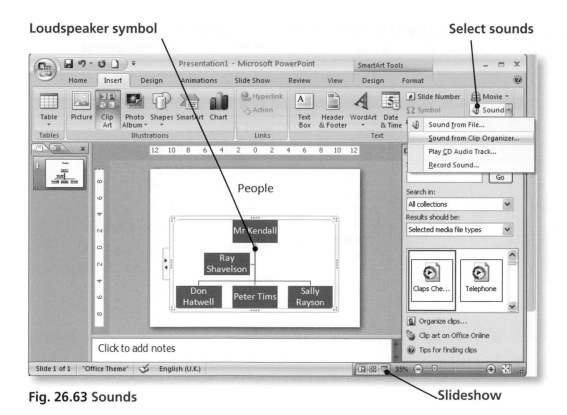

Fig. 26.63 Sounds

Slideshow

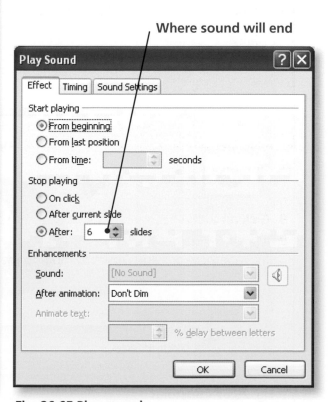

10 To remove the sound, select and delete the loudspeaker symbol.

11 To hear the sound all through a presentation, click on **Custom Animation**. Click on the down arrow for the sound in the **Custom Animation** pane and select **Effect Options**.

12 You can now start the sound on the first slide and select on which slide it will end.

Where sound will end

Fig. 26.64 Effect options

Fig. 26.65 Play sound

Insert media file

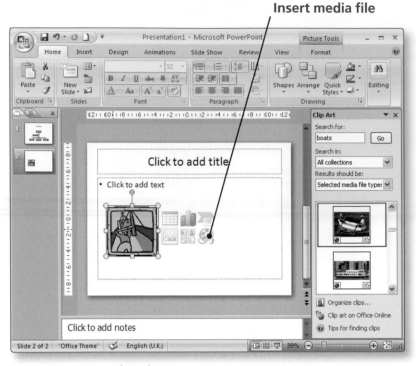

Fig. 26.66 Boat animation

Movies are normally added as moving GIF files. As with sounds, a range of animations are stored in the Clip Organiser or you can insert your own saved files.

add movies

1 Open the **Clip Organiser** and make sure you search for **Movies**, or click on this option on the **Insert** tab and look for animations on your computer or within **Clip Art**.

2 Click on an image to add it to your slide. It will look like a normal picture and can be moved and resized.

3 When you run the slide show, the 'picture' will move.

Check your understanding 14

1 Start a new presentation saved as *Christmas*.

2 Give the first slide the title *Happy Christmas*.

3 Add a new slide with the title *Enjoy!*

4 Insert a suitable movie from the **Clip Organiser** and resize it so that it fills half the slide.

5 Insert a relevant sound from the **Clip Organiser** to start automatically when the slide appears.

6 Run the slide show.

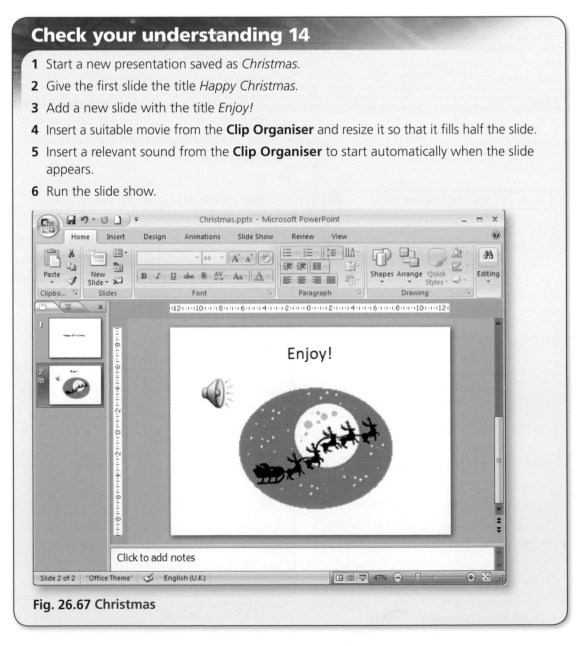

Fig. 26.67 Christmas

Animation schemes

Having decided on the main content of your slides, you can set your objects to appear or disappear as the presentation is run. A quick way is to apply animations to different types of object on the Master slide so that they are translated throughout the presentation, or you can work on each slide to select the order in which each object will appear and decide how it will behave.

set animations

1 Open the slide you want to animate.

2 Click on the **Animation** tab.

3 Click on an object such as a chart, bulleted list or title.

4 Select an option in the **Animate** box such as **Fly In** or **Fade**. Rest your mouse on different choices to preview the effect or click on the **Preview** button.

5 You can set a time for the next slide to appear in the **Advanced Slide** section.

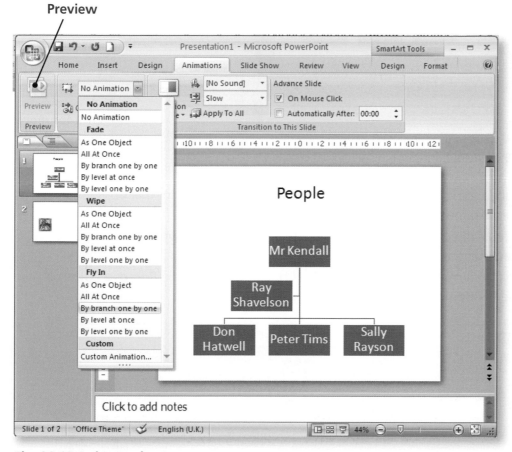

Fig. 26.68 Animate chart

6 For more detailed choices and multiple effects, click on **Custom Animation**. This opens an extra pane where you can choose:
   - the type of effect to apply for its entrance or emphasis – for example, spin, grow/shrink and so on
   - whether it will appear automatically after or with a previous animation, or only on a mouse click
   - its size and speed if relevant
   - in what order each object will appear.

7 Click on the next object and set animation details. Numbers will appear showing the order in which each object is animated. Click on the **Re-order** button to change this if necessary.

8 Click on **Play** to preview, or run the slide show to see the effect.

Order

Set to appear automatically?

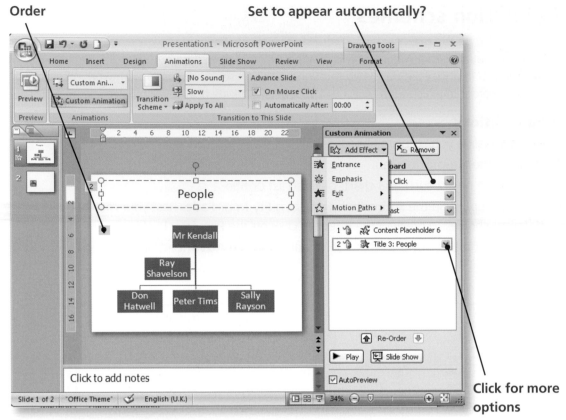

Fig. 26.69 Custom animation

9 To change some of the effects, click on the drop-down arrow for that object in the **Custom Animation** pane.

10 Click on an option such as **Timing** to open a new dialog box where you can make detailed changes.

11 Remove an animation by selecting the object and clicking on the **Remove** button or choose the **Remove** option from the drop-down arrow.

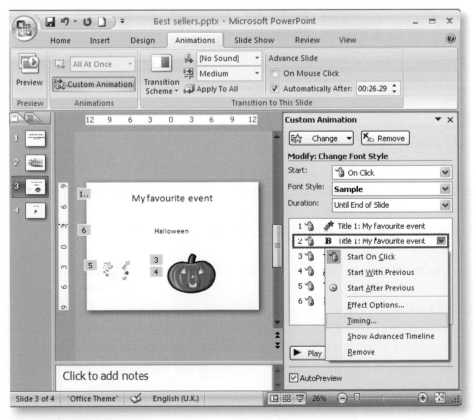

Fig. 26.70 Edit timing

Check your understanding 15

1 Open the presentation *Birds* provided on the CD-ROM accompanying this book.

2 Apply any animation to the bird image on slide 2.

3 Now apply a different animation to the slide title.

4 Run the slide show.

5 Change the order so that the title is animated first.

6 Save as *Bird animation*.

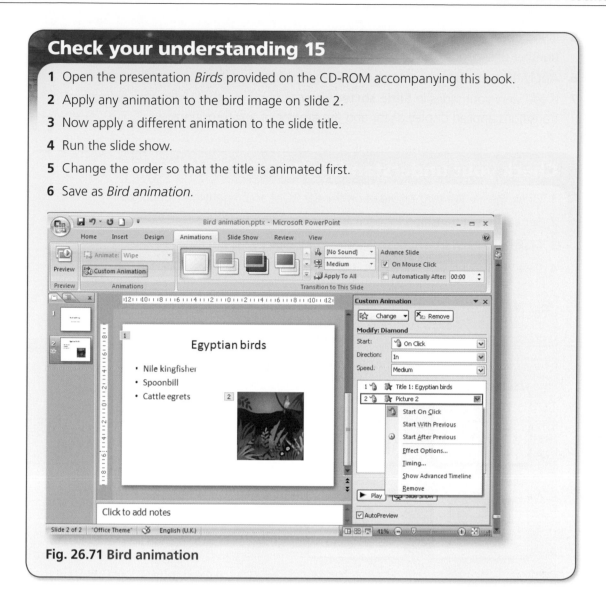

Fig. 26.71 Bird animation

Setting transitions

For a long presentation, you may like to automate the appearance of each slide or decide on a different way in which it arrives on screen. The change from one slide to the next is known as a **transition**.

set transitions

1 Click on the first slide and then click on the **Animation** tab.

2 In the **Transitions to This Slide** group, select a style of transition from the gallery. Arrows show the direction in which some effects will occur.

3 To remove a transition, click on the top **No Transition** option.

4 Set whether the transition will appear fast or slow.

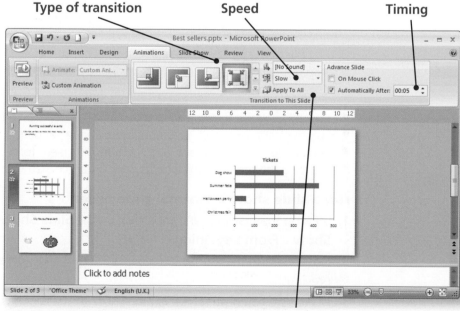

Fig. 26.72 Transition

5 If you don't want to click for each slide, deselect the mouse click option and set the number of seconds to wait before the slide will appear automatically in the **Timing** box.

6 Apply the same transition to all the slides or choose a different effect for each one.

7 If you view your slides in **Slide sorter view**, you will see that slides with animations or transitions applied display a star and the timings of any transitions.

Check your understanding 16

1 Open the file *Walking* provided on the CD-ROM accompanying this book.

2 Set a different transition for each slide.

3 Make sure each slide appears automatically six seconds after the last one.

4 Save as *Walking slides* and close the file.

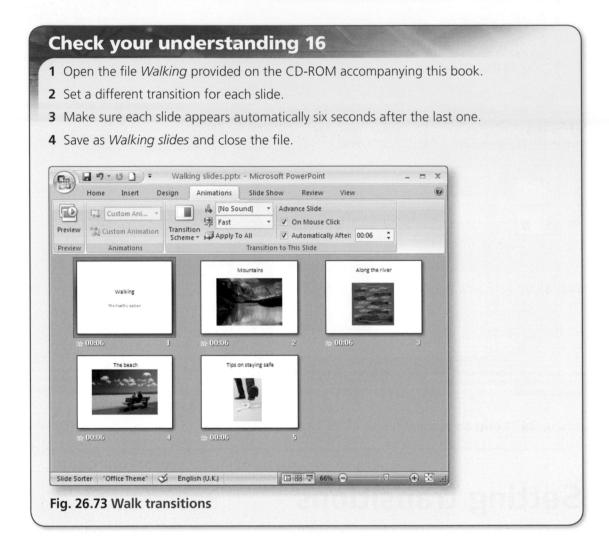

Fig. 26.73 Walk transitions

Professional slide shows

A professional presentation will not only contain slides with well set out points and suitable enhancements but will also have an introductory slide and possibly summary (a slide listing all the titles in the presentation) and conclusion slides as well. You may also need to produce accompanying documentation such as handouts containing thumbnails of all the slides.

Once your slide show is ready, you can run it in different ways. For example:

● controlled by the speaker, so that each slide appears on a mouse click
● automatically, with each slide appearing after a set period of time
● unmanned, in a continuous loop – often convenient for a conference or exhibition hall
● selecting the order in which slides appear, rather than following the original order.

run a slide show in normal slide order

1 Click on the first slide and then click on the **Slide Show view** button, or go to **Slide Show – From Beginning**.

2 Either click on the mouse to move to the next slide, or leave it to run based on timings set previously.

3 Make sure you have rehearsed the show so that timings are correct for the type of audience you expect.

check timings

1 On the **Slide Show** tab, click on **Rehearse Timings**.
2 A timing box will appear in a corner of the screen and you can see how long each slide will be displayed and how long animations take to run.
3 Click on the arrow to move on to the next object or slide, making sure you leave enough time for your audience to read any text or enjoy the effects.
4 At the end of the rehearsal, confirm the timings or click on **No** and make changes to your presentation.

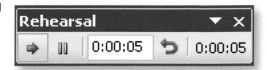

Fig. 26.74 Rehearse box

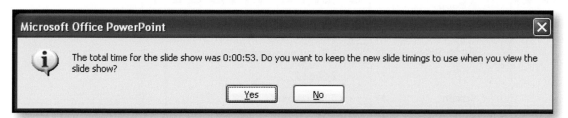

Fig. 26.75 Rehearse

run a presentation in a loop

1 On the **Slide Show** tab, click on **Set Up Slide Show**.
2 In the dialog box, click in the **Loop continuously** checkbox under **Show** options.
3 Run the slide show and stop it running by pressing the **Esc** key.
4 Take off the tick in the checkbox to remove the loop.

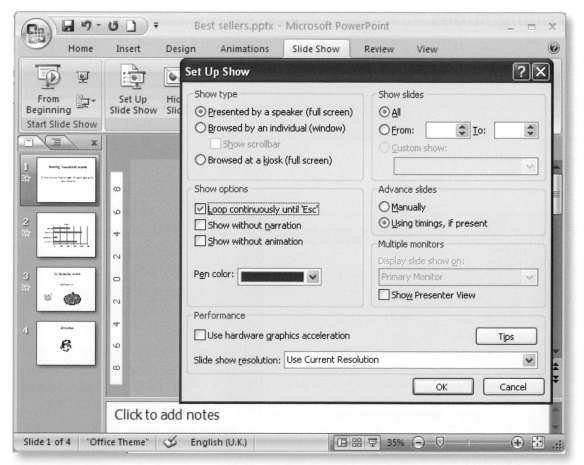

Fig. 26.76 Loop

change slide order during a slide show

1 Click on your preferred first slide before running the show, or start the show in the normal way.

2 At the point where you want to move to a different slide, right click the mouse on screen.

3 From the menu that appears, select **Next** or **Previous**, or click on **Go to Slide**.

4 All the slides in the show will be listed and you can click on the slide you now want to display.

Fig. 26.77 Go to

Hiding slides

There may be times when certain slides contain information that is not relevant to a particular audience. This slide can be hidden from view when the presentation is run, but still be available at another time.

Hidden slide

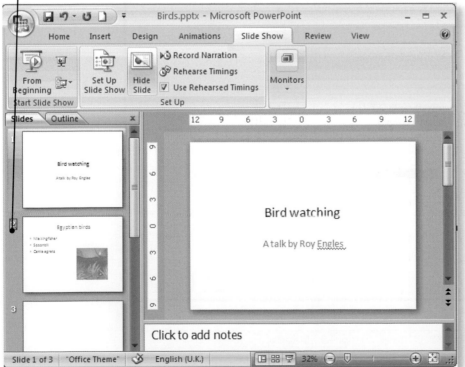

Fig. 26.78 Hide

hide a slide

1 Select the slide you want to hide.

2 On the **Slide Show** tab, click on **Hide Slide**.

3 On the **Slide** pane, or in **Slide sorter view**, its slide number will now be shown in a box.

4 When the slide show is run, hidden slides will be skipped.

5 Select the slide and click on the **Hide** command again when you want to unhide it.

Using hyperlinks

A different way to control a slide show is to embed hyperlinks in one particular slide that will act as links to different slides (or even web pages or other files). In the same way that hyperlinks work in web pages, if a slide hyperlink is clicked, it will take you to the linked slide.

You can either use text or images already present as the hyperlink, or add an action button.

create hyperlinks using slide content

1 Display the slide to contain a link and select the text or object to use as the hyperlink.
2 On the **Insert** tab in the **Links** group, click on **Hyperlink.**
3 In the **Link to:** pane, click on **Place in This Document**.
4 A list of slides will appear, so click on the slide you want to link to.
5 Click on **OK**.
6 When you run the show, hyperlink text will be coloured differently.
7 Click on the hyperlink to take you to the linked slide.

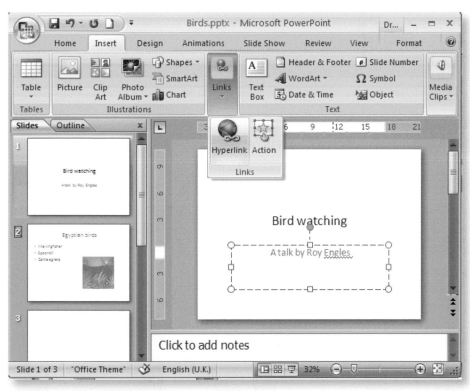

Fig. 26.79 Hyperlink1

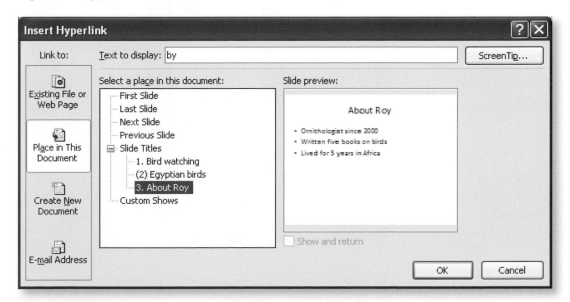

Fig. 26.80 Hyperlink2

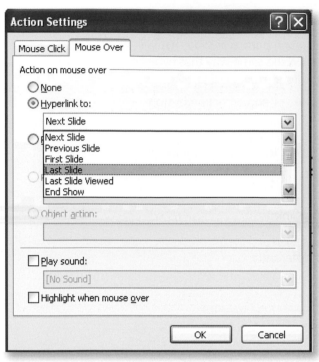

Fig. 26.81 Action

add action buttons

1 Go to **Insert – Shapes**.

2 Select an **Action** button and add it to the slide. The hyperlink window will open automatically. Or

3 Insert and format any shape you want as the button and then go to **Insert – Action**.

4 When the dialog box opens, click on **Hyperlink to:** and select which slide to display when the link is made. This could be the next or previous slide or you can click on **Slide...** and select a slide number.

5 You can choose a mouse click or hover to activate the hyperlink, and whether the shape should be highlighted when the mouse moves to it.

6 Click on **OK**.

Check your understanding 17

1 Open the file *Letters* provided on the CD-ROM accompanying this book.

2 Animate the image on slide 1.

3 Add a link from slide 2 to slide 5 showing a chart, using the text *range of magazines* as the link text.

4 Run the presentation as a looped presentation.

5 Take off the loop and now start a normal slide show. Use the hyperlink to move from slide 2 to slide 5 before returning to slide 2 again to continue in the normal way.

6 Add transitions and make each slide appear after a five-second period.

7 Finally, hide slide 2 and check that it doesn't appear when you run the slide show.

8 Save as *Letters amended*.

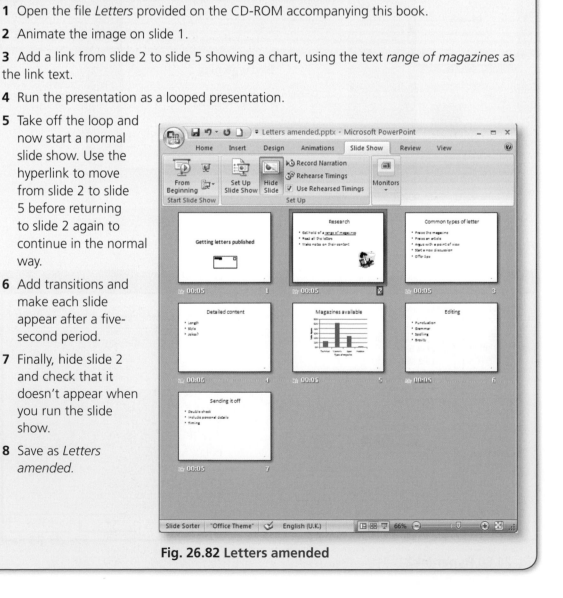

Fig. 26.82 Letters amended

Saving a presentation for use elsewhere

If you want to run a slide show on a computer that does not have PowerPoint 2007, it is possible to save it with software known as a viewer that will still enable it to be run. When you copy your Microsoft Office PowerPoint 2007 presentation to a CD, a network, or a local disk drive on your computer, Microsoft Office PowerPoint Viewer 2007 and any linked files (such as movies or sounds) are copied as well. (Formerly you had to make use of a facility known as 'pack and go'.)

You can also save your file in a form that will open directly into a slide show, rather than **Normal view**.

save a presentation to a CD that will include the viewer

1 Click on the **Office** button.

2 Select **Publish**.

3 Click on **Package for CD**.

4 Click on **OK** when a warning appears, as some files may need to be updated to run in the viewer.

5 Now name the CD and copy the files across.

Fig. 26.83 Package warning

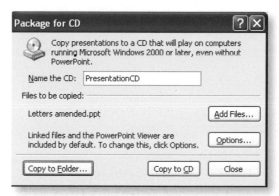

Fig. 26.84 Copy to CD

save a presentation as a show

1 Click on the **Office** button.

2 Select **Save As** and select the **PowerPoint Show** option.

3 Name and save the file in the normal way.

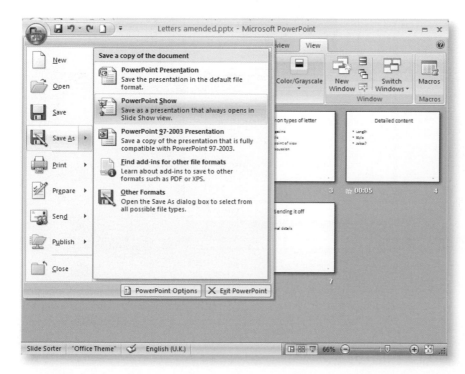

Fig. 26.85 Save show

Printing a presentation

You can print a number of objects from a PowerPoint file, including:

- all or selected slides
- handouts to accompany a talk
- speaker's notes
- the outline of the presentation.

Depending on what you want to print, you need to make sure the slide or page is set up correctly. Layout changes can be made when viewing the presentation in print preview or from commands on the **Design** tab.

create speaker's notes

1 Select any slide for which to make notes and, in **Normal view**, type your notes into the **Notes** pane.

   Or

2 Go to **View – Notes Page**. This view will show you how the page will appear when printed.

3 Type your notes into the text box presented below the thumbnail image of the slide.

4 If you need more room for the text or want the slide image larger, just drag the borders of the relevant box.

5 Repeat for any other slides where notes are required and then save the presentation.

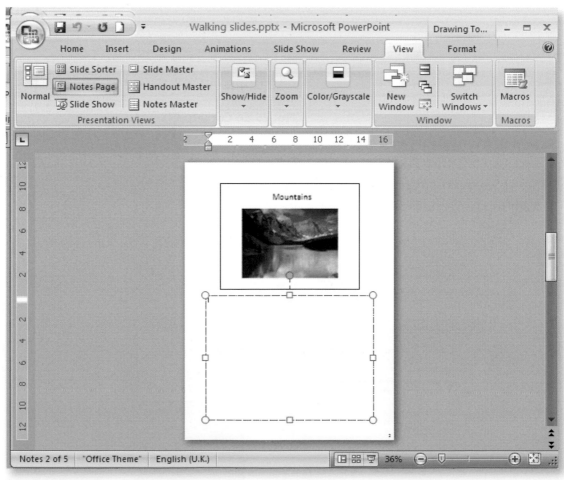

Fig. 26.86 Notes

make changes from print preview

1 Click on the **Office** button.

2 Select **Print – Print Preview**.

3 Click on **Options** if you want to print hidden slides, add or amend headers or footers or add a frame/border round each thumbnail.

4 You can also click on the **Colour/Greyscale** button to view how a coloured presentation will appear if printed in black and white.

5 Click on **Orientation** to change from Portrait to Landscape.

6 Click on **Zoom** to zoom in or out to check slide details or the overall effect.

7 Change the content of the **Print What:** box to select slides, handouts and so on or set how many thumbnails to a page.

8 Move through a presentation by clicking on the **Next Page** command.

9 Print the presentation by clicking on the **Print** command.

10 Return to normal view by clicking on the **Close** button.

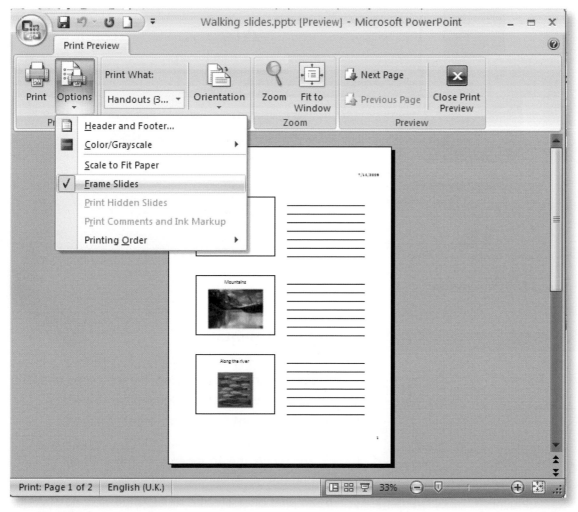

Fig. 26.87 Print preview

change layout from the ribbon

1 Click on the **Design** tab.

2 Click on the **Slide Orientation** command to change slide orientation from Landscape to Portrait.

3 Click on **Page Setup** to open the dialog box. You can now select the type of printout you want and its orientation and size.

Fig. 26.88 Page setup

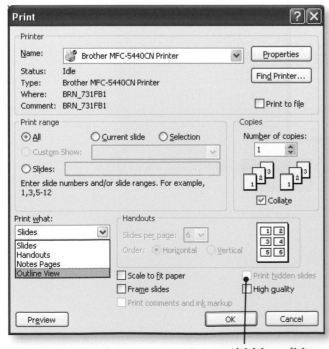

Fig. 26.89 Print box   **Print any hidden slides**

print a presentation

1 Click on the **Office** button.

2 Go to **Print – Quick Print** to print a copy of all the slides, one per page, using default settings.

3 Click on **Print** to open the **Print** dialog box.

4 Here, select the type of printout, colour or black and white, the arrangement of thumbnails or number of copies you want.

5 To print only selected slides, enter slide numbers into the **Print range:** box or select the slides first – for example, in **Slide sorter view** by holding down **Ctrl** as you click on each in turn. Then click on the **Selection** option.

6 Click on **Preferences** to change your printer's settings.

7 If necessary, click on in the checkbox to print hidden slides.

8 Click on **OK** to print.

Check your understanding 18

Check your understanding 18

1 Open the file *Letters.*

2 Print handouts showing all the slides, four slides per page.

3 Print a copy of slides 1 and 3 only.

4 Finally, add the following note for slide 2: *Mention library opening times*. Print a copy of just this page.

5 Save the file as *Letter notes*.

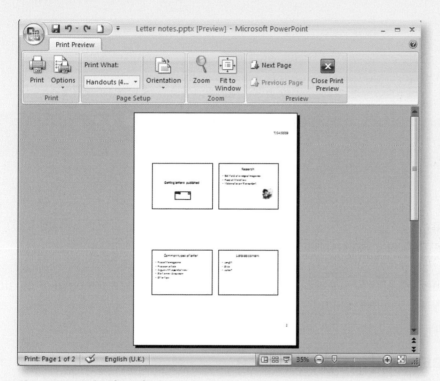

Fig. 26.90 Print handouts

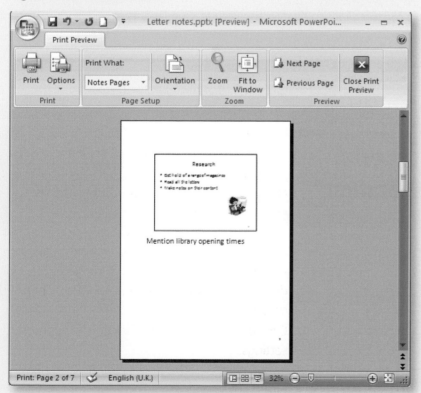

Fig. 26.91 Print notes

Assignment

This practice assignment is made up of four tasks

- Task A – Set up a presentation template
- Task B – Create a presentation
- Task C – Animate a presentation
- Task D – Edit a presentation and print

You will need the following files:

- About C&G.doc
- Assessment.doc
- Contents.doc
- Course structures.doc
- Level 2 units.doc
- Summary.doc
- C&G dog logo.jpg
- C&G Guidance website.jpg
- C&G logo.jpg
- C&G website.jpg
- Intro.mp3

Scenario

You work in an office and have been attending a City & Guilds e-Quals evening course which has noticeably improved your work performance. e–Quals is an IT qualification covering many of the skills used in ITQ.

Your line manager has been so impressed, that they have allocated you some time to produce a presentation to run on an endless loop in reception to promote this excellent qualification.

Please read the text carefully and complete the tasks in the order given.

Task A – Set up a presentation template

1   Start a presentation graphics application such as PowerPoint and select a simple template.

2   Edit both the **Title Master** and **Slide Master** slides to change the default slide layouts to a style of your choice.

Include graphic **C&G logo.jpg** for the City & Guilds logo.

Add a footer with page numbering, date and your name.

Make any other changes to the background and text attributes to your taste.

Take screen shots of these master slides being edited.

3   Save the new template as **C&G** into an appropriate location and close the application.

Take a screen shot of the template being saved.

Task B – Create a presentation

1   Start a new presentation based upon **C&G** template you created in task A.

Create a title page with this text:

Copy A please supply

Copy B please supply

Include the graphic **C&G dog logo.jpg**

2   Create a contents page with this text:

Contents

- About City & Guilds
- Courses structures
- Level 2 Unit Choices
- City & Guilds Websites
- Assessment

The text is in a document named **Contents.doc**. You should open this document then copy and paste the content into your presentation slide.

Format and arrange the slide to your taste.

3   Create a page about City & Guilds using the text in a document named **About C&G.doc**. You should open this document then copy and paste the content into your presentation slide using **Outline view**.

The text under the second heading, City & Guilds Group, should be copied and pasted from page 2 of this web page:

Format and arrange the slide to your taste.

4   Create a page about the City & Guilds courses structures using the text in a document named **Course structures.doc**.

You should open this document then copy and paste the content into the boxes of an organisational chart on this slide to show the structure of the courses.

Format and arrange the slide to your taste.

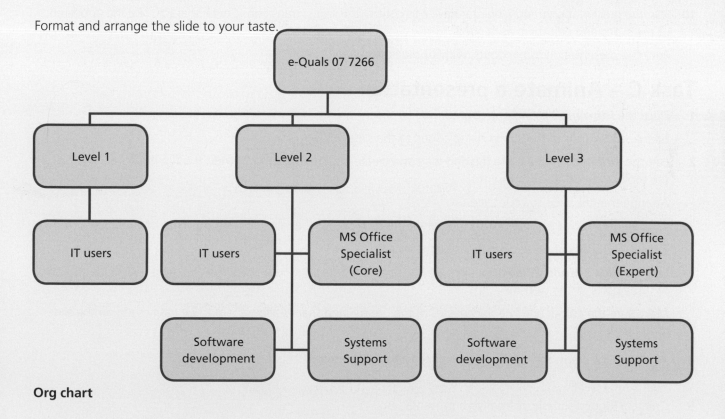

Org chart

5   Create a page about unit choices using the text in a document named.

You should open this document then copy and paste the content into your presentation slide using a (2x15) table to hold the unit numbers and titles.

should be in the first row of the table. As it is a core unit, set the font to bold for this row.

Format and arrange the slide to your taste.

6   Create a page about City & Guilds websites entitled **City & Guilds Websites**.

Use the graphic named **C&G website.jpg** with a text box close to it to show the web address: www.cityandguilds.com

Use the graphic named **C&G Guidance website.jpg** with a text box close to it to show the web address: www.cityandguilds.com/15998.html

You may need to resize these graphics. Group each graphic with its text box.

Format and arrange the slide to your taste.

7   Create a page about assessment using the text in a document named **Assessment.doc**. You should open this document then copy and paste the content into your presentation slide using **Outline view**.

There should be three subheadings on this page with the rest of the text, set as bullet points.

Format and arrange the slide to your taste.

8   Create a summary page by copying the contents page, editing it into:

Summary

- About City & Guilds
- Courses structures
- Level 2 Unit Choices
- City & Guilds Websites
- Assessment

9   Check the text in your presentation for spelling, adding to the spell checker dictionary where appropriate.

10   Run the presentation using a pointer device to control the slides' transition. Check that you like the appearance of the slides and edit them as needed.

Print the presentation as handouts, with six slides per page.

Task C – Animate a presentation

1   Place the **Intro.mp3** sound object onto the title slide.

Take a screen shot of the sound being added to the slide.

2   Edit the **Title** slide to duplicate the dog logo and create a mirror image of it. (Add the flip horizontal button to the toolbar, if not already present.)

Arrange the title slide to your taste.

Animate the dog logos and text so that when the slide opens, they slide in from the sides of the screen.

Take screen shots of these animations being set.

3   Edit the **City & Guilds Websites** slide, adding some callouts from auto shapes with simple text to identify two or three parts of the website.

Make sure the colour and line thicknesses of the pre-defined shapes allow them to be easily seen on the slide.

Print just this slide.

4   Arrange a different transition action between each of the slides.

Take a screen shot of a transition action being set.

5   Create a looped slideshow from your presentation.

Test it until you have established an appropriate length of time for each slide to appear.

Take a screen shot of the looped slideshow being set.

Task D – Edit a presentation and print

1   Carefully view your presentation then perform a final edit on it.

Use graphical text on a slide where you think it most appropriate.

Activate the ruler and guidelines then check the position and alignment of your text and graphical objects:

a)   Dynamically using the ruler
b)   By setting positional properties

2   Print the whole presentation as notes pages.

3   Save your presentation as a slideshow.

Take a screen shot of it being saved.

UNIT PIM

Personal information management software

This unit involves the use of the **Personal Information Management (PIM)** program Microsoft Outlook 2007. You will learn how to receive and send email messages including attachments, how to organise messages and how to use other aspects of the software to create and use the address book, set diary events and create notes and to-do tasks.

At the end of this unit you will be able to:

- navigate and use Outlook facilities
- use Outlook on a network to communicate
- use the Outlook calendar
- use Contacts, Tasks and Notes.

Opening and closing Outlook

As with all Microsoft Office programs, you launch the program from the desktop, **Start** menu or **All Programs** list and exit using the file menu or shortcut **Close** button in the main window.

launch Outlook

1 Click on an icon or first open the **Start** menu.
   Or
2 Go to **Start – All Programs – Microsoft Office – Microsoft Office Outlook 2007**.

close Outlook

1 Go to **File – Exit**.
   Or
2 Click on the **Close** button in the top right-hand corner showing a cross.

Fig. 31.1 Launch Outlook

Outlook structure

Unlike simple email systems such as Outlook Express, you can do much more than send or receive messages and build up an address list (Contacts) when using Outlook 2007. Other activities include:

- organising events and maintaining a diary using the **Calendar**
- monitoring actions over time using the **Journal**
- creating and delegating a variety of **Tasks**
- keeping important information or actions in mind by creating **Notes**.

When you open Outlook, you will see a number of panes related to the activities that can be performed. Some of these panes can be expanded or collapsed by clicking on the double arrows in the corner.

When you are in a particular section such as Mail or Calendar, its name will be highlighted in orange in the Navigation pane. Move to a different section by clicking on the **heading** or **button** in the pane or select the section from the **Go** menu. When you do this, you will find that different, relevant toolbar buttons have now been made available, although the seven main menus are always visible.

The contents of any selected folder will be displayed in the **main pane** and the folder's name will be highlighted in light grey in the **Navigation** pane. Other panes such as a **Reading** pane can be set to appear if required.

In some views, you will see a list of all the folders within Outlook. If this is not visible, click on the **Folder List** button to display it again.

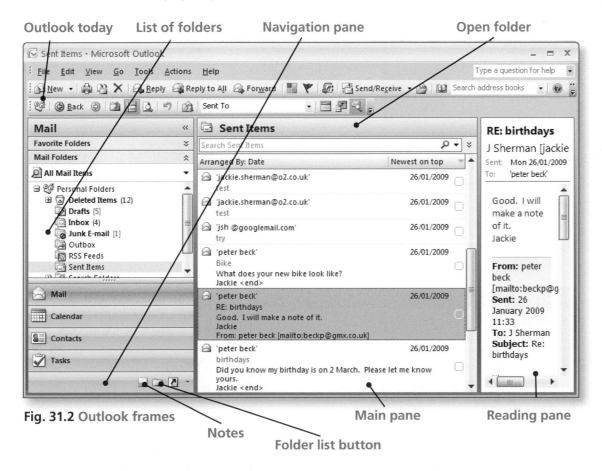

Fig. 31.2 Outlook frames

Labels: Outlook today · List of folders · Navigation pane · Open folder · Notes · Folder list button · Main pane · Reading pane

Customising views

With so many ways of viewing messages, contacts or appointments, you may want to customise the way Outlook displays information. To do this, you need to make changes from the **View** menu or toolbar commands.

If you click on the **Outlook Today** command, you can see a summary of any appointments, tasks or messages for the current week.

customise Outlook views

1 Click on the **View** menu.

2 Click on one of the items such as **Navigation** pane or **Autopreview** (which offers some of the message text) to turn it on or off or change its position.

3 Click on **Arrange By** to sort messages or contacts in the main pane on various criteria such as date or subject.

4 Click on **Current View** to see the present arrangement (it will show a tick) and select alternative ways to display items in the main pane.

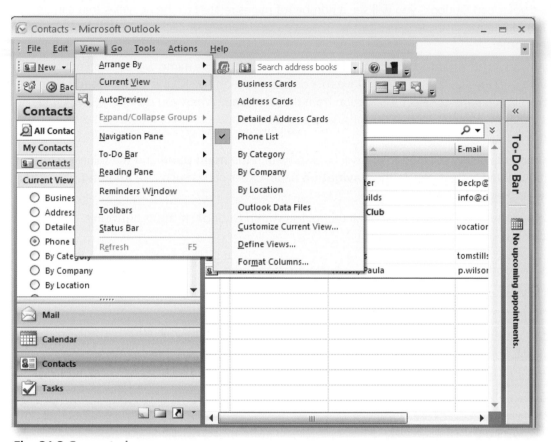

Fig. 31.3 Current view

5 Click on **Current View – Customize Current View** to make more detailed changes. For example, for messages or contacts, click on **Fields** to add or remove the headings under

Fig. 31.4 Show fields

which message or address details will be displayed.

a To add a heading, click on it in the left pane and click on **Add**.

b To remove a heading, click on it in the right pane and click on **Remove**.

c To change the order of headings across the main pane, move headings up or down the list using the **Move** buttons.

6 Shortcuts to many of these options are available from the toolbar.

Reading pane   Fields

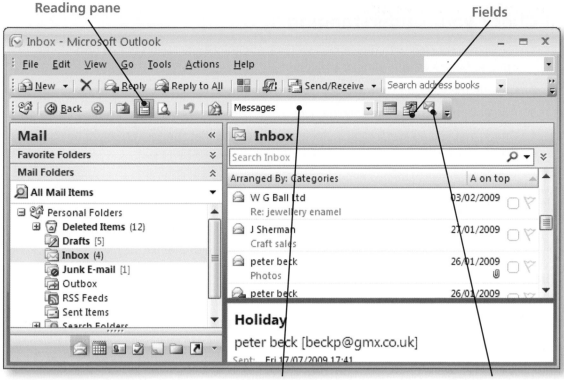

Fig. 31.5 View buttons   **Current view options**   **Autopreview**

Note that in Outlook 2007 you can even create your own floating toolbars.

create a toolbar

1 Right click on any empty area of a toolbar and select **Customize**.
2 On the Toolbars tab click on **New** and enter a name for the toolbar.
3 It will now appear floating in the main window.
4 Click on the **Commands** tab and drag any buttons you want on your toolbar straight onto the bar.

Action menu

It is worth noting that shortcuts to many of the actions you can perform in Outlook 2007 are available from the **Actions** menu. You should check here for a quick way to carry out a task, which will depend on the part of the system you are currently working with.

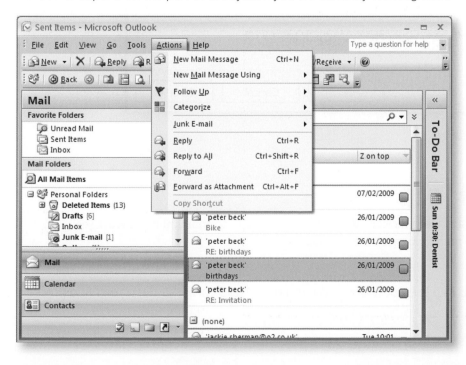

Fig. 31.6 Action menu

Check your understanding 1

1 Familiarise yourself with the various sections of Outlook, including the **Action** menu options for the various parts of the system.

2 Change the way you can view Contacts.

3 Add and then remove the Reading pane.

4 Check that you are organising incoming messages so that they are arranged by date.

Filtering

If you only want to see messages from a particular person, or about a single subject, you can filter out all other messages temporarily and just display those of interest.

filter a view

1 Go to **View – Current View – Customize Current View – Filter**.

2 When the filter window opens, enter the search criteria such as who the messages are from or their subject.

3 Click on **OK**.

4 You should now only see relevant messages in the main window.

5 Clear the filter by clicking on **Clear All**.

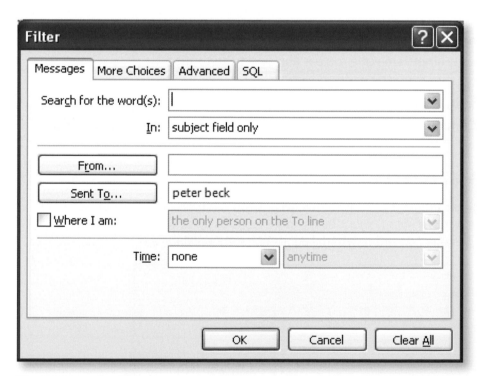

Fig. 31.7 c and g filter

Using Help

Outlook has the usual **Help** facilities built into the program, so that you can always see demonstrations or read guidance on how to carry out a task.

use Help

1 Type key words into the query box in the top right-hand corner of the screen and press **Enter**.

Or

2 Open the **Help** menu and select **Microsoft Office Outlook Help**.
Or

3 Press the function key **F1** or click on the **question mark** in a blue circle at the end of the **Standard** toolbar.

4 This opens the **Help** pane where you can follow up a relevant topic heading.

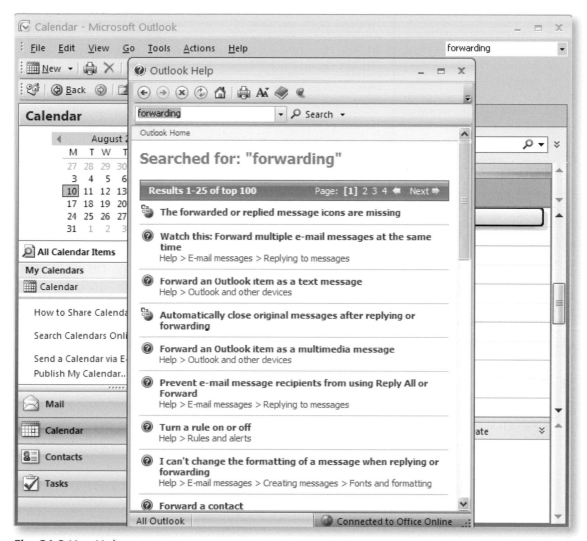

Fig. 31.8 Use Help

Check your understanding 2

1 Use **Help** to find out about **Notes** in Outlook 2007.

2 Now use **Help** to find out the difference between **Reply** and **Reply to All** when replying to emails.

Creating new messages

The **New** button is a shortcut to creating a new object relevant to the section of Outlook you are currently working with. If you want to use it to create a new email message and have been working in Calendar or Contacts view, you need to open the **New** menu from the drop-down arrow and select the correct object.

Fig. 31.9 New options

create a message

1 Click on the **New** button or click on the drop-down arrow and select the **Mail Message** option if you are not in Mail.

2 When a new message window opens, fill in the details correctly.
  - The **To:** box will contain the full email address of one or more recipients (separate them with a semicolon).
  - The **Cc:** box will contain the full email address of people receiving a copy.
  - The **Subject:** box will contain a summary of what the message is about.
  - The main window contains your message.

3 After checking for errors, click on the **Send** button to send the message.

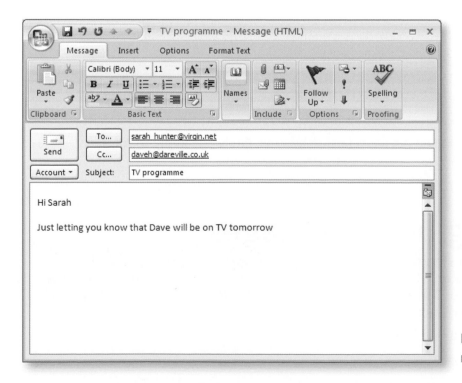

Fig. 31.10 Create message

Blind copies

To send a copy of a message to someone confidentially – i.e. without the knowledge of the other recipients – you need to add their email address in a **blind copy** or **Bcc:** box.

send a blind copy

1 Click on **Options**.

2 Click on **Show Blind Copy**.

3 Add the address details to the new box that will appear.

Send blind copy

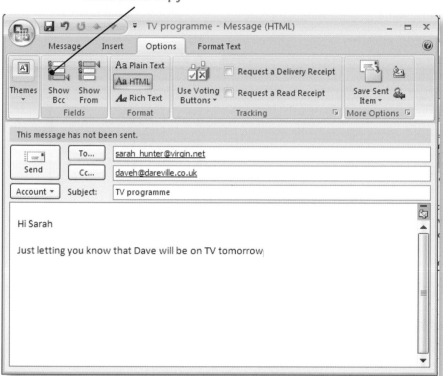

Fig. 31.11 bcc

Check your understanding 3

1 Start a new message.

2 Address it to a friend or colleague who has given you their email address and is happy to send and receive practice messages.

3 Send them the following message:

   a Subject – Working with email

   b Message – Hi, here is a practice email.

4 Send a copy to yourself by adding your full email address in the **Cc:** box.

5 Send the message.

Recalling messages

If you send a message by mistake (for example, you forget to attach a file before sending), you may be able to recall it and resend it later. For this, you and the recipient must both be using Microsoft Exchange.

recall a message

1 Open your **Sent Items** folder.

2 Open the message.

3 Click on **Other Actions** and select the **Recall** option.

4 You can choose whether to delete unread messages or replace them with a new message.

5 You can also choose whether to be notified that the recall has taken place.

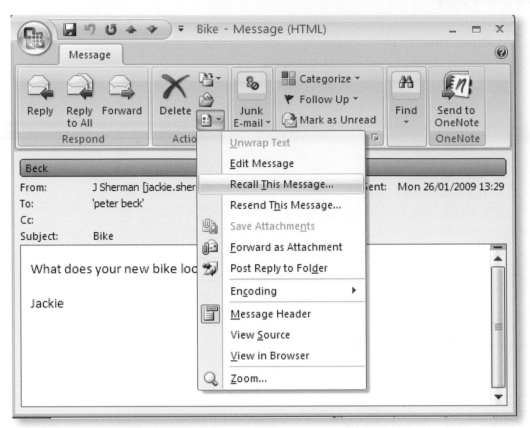

Fig. 31.12 c & g recall message

Email signatures

If you use the same email address for work, leisure or other aspects of life, you may be writing different types of messages and need to sign these in different ways. You may also need to add details such as job title, phone numbers, addresses and so on to messages that will take time to repeat for new messages. To add certain details to the ends of your messages automatically, you can create one or more signatures. These can then be used as appropriate.

create email signatures

1 To open the **signature** dialog box from the main Outlook window, go to **Tools – Options**, click on the **Mail Format** tab and then click on the **Signatures** button.

2 When creating a message, click on **Insert – Signature – Signatures**.

3 In the window that opens, click on **New**.

4 Any signatures created previously will be listed and you can select any to make changes, rename or delete them.

5 Give your new signature a name so you can identify it in future.

6 Type the details into the **Edit** window and format them as you want them to appear in future messages.

7 Click on **Save**.

8 Click on **OK** to return to your message.

9 If you will want to add one particular signature to all your messages, you can set the named signature in the **New messages:** box so that it is added automatically.

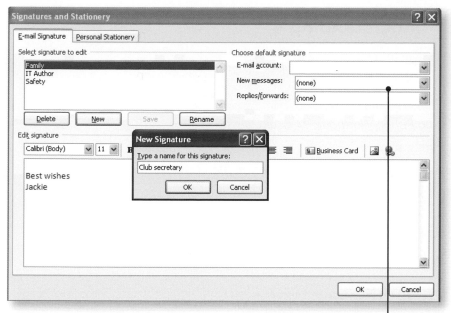

Fig. 31.13 Signature1   Add by default

add a signature to a message

1 Type the message as normal.
2 Click at the end of the main text.
3 Click on **Insert – Signature**.
4 Select the correct details from the list.
5 They will appear in your message.

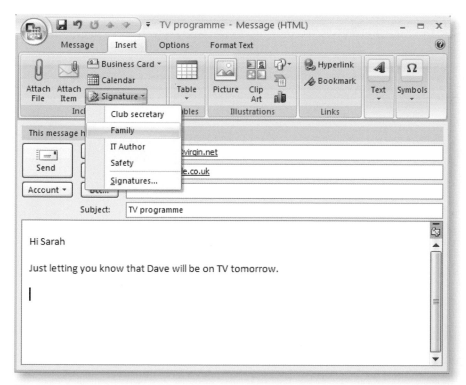

Fig. 31.14 Signature2

edit a signature

1 Open the signature dialog box.
2 Select the signature you want to change in the top pane.
3 Make your changes in the main window.
4 Click on **Save**.
5 Click on **OK**.

Check your understanding 4

1   Start a new message to send to a friend.

2   Give it the subject *Museum*.

3   Create a signature with the name *Curator*.

4   Add the details *Jo Bishop, Northome Museum, Sussex.*

5   Return to your message.

6   Write the message *The museum will now close on Mondays.*

7   Now make the following change to your signature: the county should be *Shropshire*, not *Sussex*.

8   Add your *Curator* signature to your message.

9   Send the message.

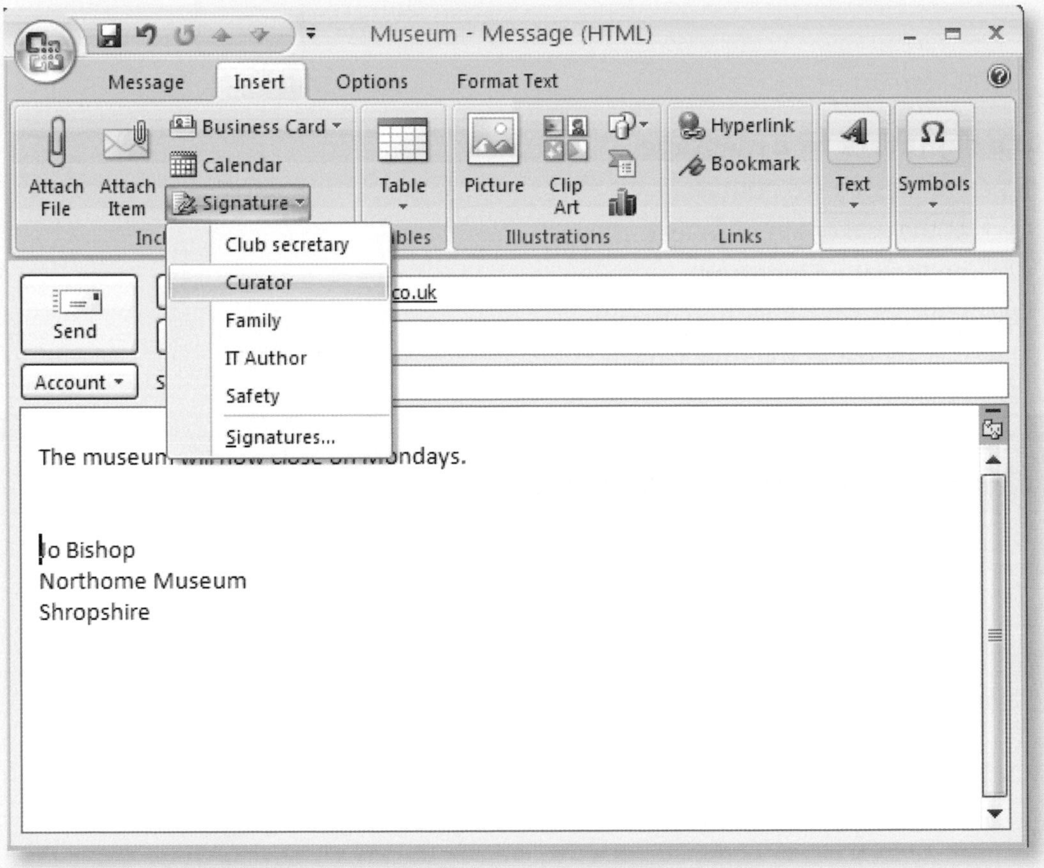

Fig. 31.15 Add signature

Attaching files

It is quick and easy to add files from your computer to a message.

add attachments

1   Start a message.

2   Click on the **Attach File** button showing a paperclip.

3   Browse through your folders until you find the first file to attach.

4   Click on its name and then click on **Insert**.

5   Repeat to add further files.

6   They will be listed in a new **Attached** box.

7   Send the message as normal.

Attach button   Attachments

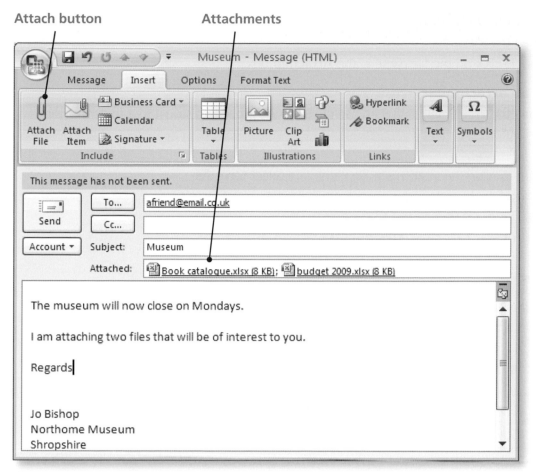

Fig. 31.16 Attachments

Check your understanding 5

1 Start a new message.

2 Address it to a friend and send a copy to yourself.

3 Attach the image file *Party* provided on the CD-ROM accompanying this book.

4 Give the message the subject *Invitation*.

5 Add the following text: *Steve, Hope you can come to the party on Saturday. See you.*

6 Sign and send the message together with its attachment.

Receiving attachments

When a message has an attachment, you can do a number of things:

- open and read/view the file
- save the attachment outside Outlook
- print the attachment
- delete the attachment.

As attachments are known as one way to spread viruses, you will see a warning message when you try to open an attachment. As long as you are happy that it is genuine, you can open it. Otherwise, click on the **Save** option to save the file and scan it with a virus checker before opening.

Fig. 31.17 Email warning

open an attachment

1 Open the message.

2 Double click on the file name in the **Attached** box.

save an attachment

1 Open the attachment and then use the **Save As** process to save it to a folder on your computer.

Or

2 Right click on the attachment file name and select **Save As**.

Or (useful where there are several attachments)

3 Click on the **Office** button.

4 Select **Save As – Save Attachments**.

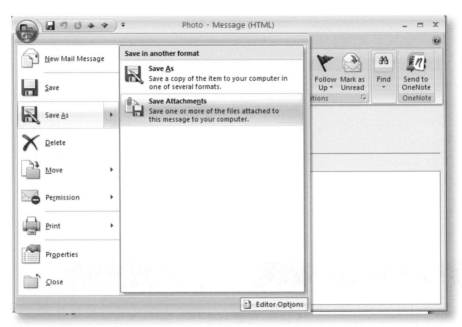

Fig. 31.18 Save attachment

5 Select a folder in which to save the file(s) and save as normal.

Fig. 31.19 Save attachments 2

delete an attachment

1 Right click on the file name in the **message** box.

2 Click on **Remove**.

> ## Check your understanding 6
>
> 1 Open the message *Invitation* you sent to yourself earlier.
> 2 Save the image file attached with the new filename *Party time* to a folder on your computer.

Compressing attachments

Some files can be very large to send and so it is a good idea to compress them. This reduces their size so that they are sent more quickly and are easier to receive. This process is known as **zipping** or **archiving** and a program to do this is built into Windows XP or later machines. Once zipped, the archive can be treated exactly like a normal file and attached to an email message.

compress a file

1 Right click on the file on the desktop, or select several files at the same time and right click on any one.

2 Select the option to **Send to – Compressed (zipped) Folder**.

3 A yellow archive folder displaying a zip will appear. This will have the same name as a selected file and will contain the files in a compressed format.

4 Rename the archive if necessary.

5 Attach it to an email message in one of the following ways:
 • Right click on the archive and select **Send to – Mail Recipient**. This will open a new email message with the archive already attached.
 • Create a new message from within Outlook and attach the file as described above.

Fig. 31.20 Compress

Compressing picture files

When you try to send an image file using email from the desktop, you will be asked if you want to compress it first, so that it takes up less room. Only click on **OK** if the picture resolution is not important.

Fig. 31.21 Compress picture files

open a compressed file

1 When a message containing an archive arrives, open the message and then double click on the attachment.

2 Either double click on any files it contains to open them individually on screen or select all or some of the contents of the archive and extract them using the wizard. Do this by clicking on the **Extract** option and working through the **Extraction Wizard** step by step.

3 Before they are extracted and stored, you will need to browse through your folders to choose where the files will be extracted and saved to.

Open wizard

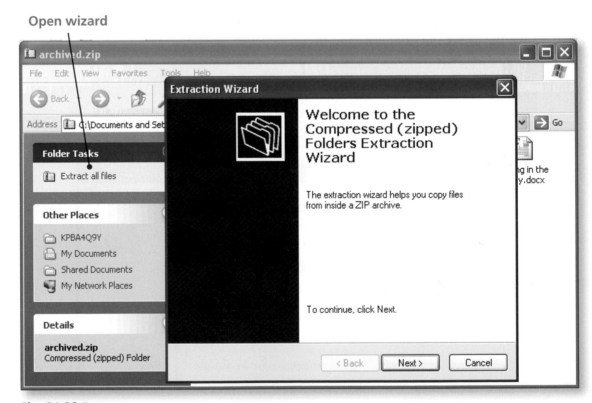

Fig. 31.22 Extract

Check your understanding 7

1 Practise compressing two or more files on your desktop – for example, in My Documents.

2 Use the wizard to extract them from the zipped archive to a different folder location.

Formatting messages

When you create messages in HTML code or Rich Text Format rather than plain text, you can format your messages and view formatting in messages you receive.

set message format

1 To set the formatting for all future messages go to **Tools – Options**.

2 Click on the **Mail Format** tab.

3 Select your preferred option from the drop-down list in the window.

4 Click on **OK**.

5 When composing, you can change the format by clicking an alternative on the Options tab. If you move to Plain Text, you may be warned that some features will be lost.

After creating a message, click on the **Format Text** tab to see the familiar formatting tools such as Bold, Underline, font type and size boxes. Select and apply them to your text in the normal way.

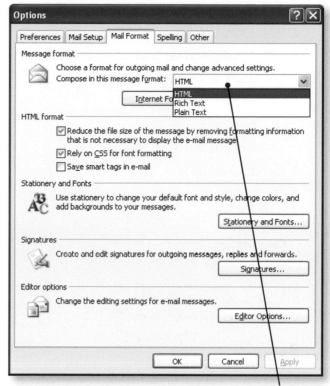

Fig. 31.23 Mail format   **Set format**

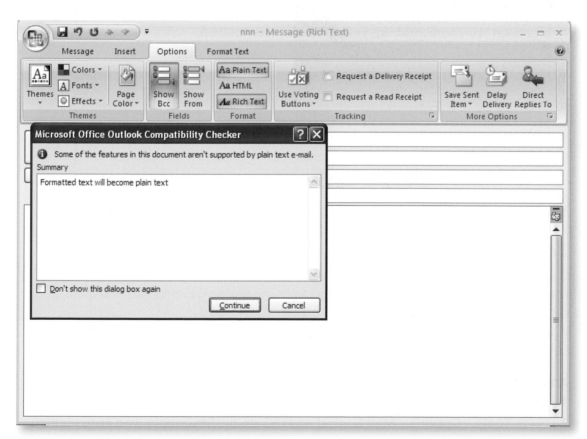

Fig. 31.24 Message format when composing

233

Check your understanding 8

1 Compose a new message addressed to yourself.

2 Add the subject *Formatting*.

3 Add the message *I want to check that my text has been formatted.*

4 Make sure you are using an appropriate message format.

5 Click on the **Format Text** tab and format the text as follows:

    a '**want to check**' bold

    b '<u>my text</u>' underlined

    c 'been' red

    d 'formatted' – larger font size.

6 Send the message.

7 To view the formatting, open the **Inbox** and double click on the **Formatting** message when it arrives.

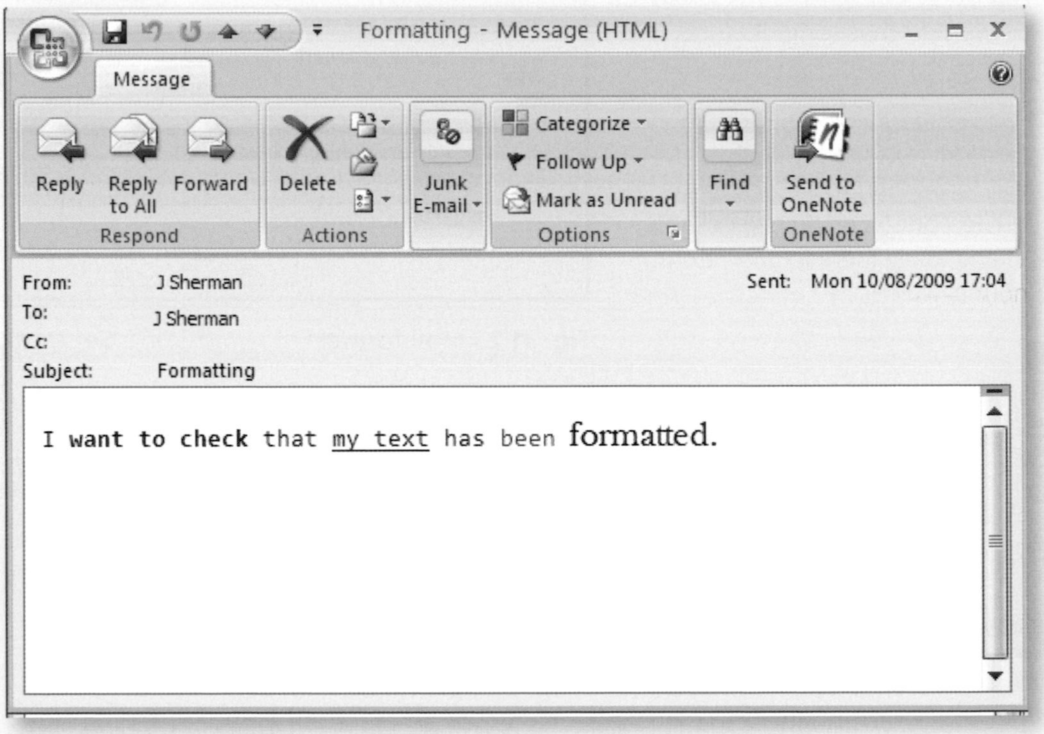

Fig. 31.25 Formatted

Templates and themes

Outlook provides a range of backgrounds and associated font styles and colours known as **themes** or **stationery** that can be applied to make messages more attractive, and these are kept in the stationery section. They only apply to HTML messages.

If you want to use themes regularly, or repeat a message that has been carefully formatted, you can save messages as ready-prepared templates that can be used again many times.

apply a theme or stationery

1 Make sure HTML is the selected message format.

2 Go to **Tools – Options – Mail Format** and click on the **Stationery and Fonts** button.

3 On the **Personal Stationery** tab, click on **Theme**.

4 Browse through the list and select your preferred style.

5 Click on **OK**.

Next time you compose a message, the theme or stationery you chose will be applied.

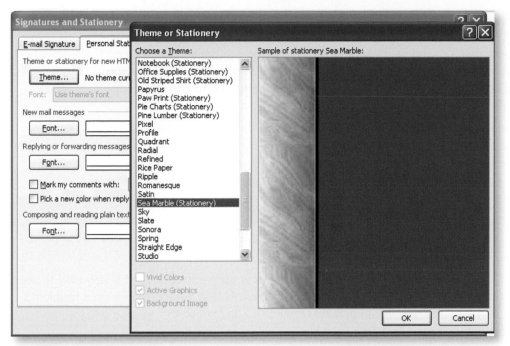

Fig. 31.26 Stationery

use templates

1 Create and format the basic message you will want to use as your template.

2 Click on the **Office** button.

3 Click on **Save As**.

4 Select **Outlook Template** in the file type box and save the template to the default folder with a recognisable name.

5 When you want to create a new message based on the template, go to **Tools – Forms – Choose a Form**.

6 Click on **User Templates in File System**.

7 Select the template from the list and click on **Open**.

8 Create your new message as normal.

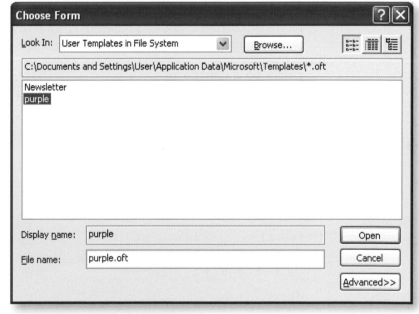

Fig. 31.27 c and g use template

Check your understanding 9

1 Create a new message with stationery of your choice as the background.

2 Save it as a template named *My stationery*.

3 Create a new message based on this template and send it to yourself.

4 When it arrives in your Inbox, check that it still displays the chosen background.

Replying and forwarding

When you receive a message in your Inbox, you can reply to the sender by using the automatic reply facilities. You can also send the message on to a third person using the forwarding facilities. When you reply, attachments should not normally be returned to the sender, but they will be sent on automatically when you forward messages.

Note that there are some email messages that you should not reply to or even open. These may involve chain letters, disguised rogue programs or phishing emails that try to entice you to waste money, open attachments that then download harmful material or give away personal details. Whenever you receive a message from an unknown source, always be wary and take sensible measures to protect yourself and your machine.

Because email is a faceless form of communication, it is easy to misunderstand what has been written. If you are unsure about an email, sometimes the best solution is to speak to the person who sent it, either on the phone or face-to-face.

Misunderstandings are not the only problem in electronic communication – it is very easy to send an email but forget the attachment, or to load attachments that are too big for your recipient's inbox. If you make a mistake, it can be easily remedied - just remember to send a follow-up email explaining what your next step will be.

As a recipient, there are hazards of IT-based communication to be aware of. Spam is a fact of life these days, but it is not the only form of malicious incoming email. Email hazards range from phishing emails to inappropriate content, from viruses and spyware to hidden key-logging software. The best way to be safe is to use your head – don't click on links or download attachments from unknown senders, and think twice about clicking on a suspicious email from a trusted contact – they could have fallen victim to a virus or trogan.

reply to a message

1 Select the message and right click or click on the **Reply** toolbar button to reply only to the sender.

2 Click on **Reply to All** if you want everyone listed in the **Cc:** box to receive a copy of your reply as well.

3 When the reply message window opens, you will see entries in the **To:** and **Subject:** boxes already completed. In the **Subject:** box, the original subject text will be preceded by **Re:**.

Reply to everyone

Reply to sender

Completed boxes

New text

Original message details

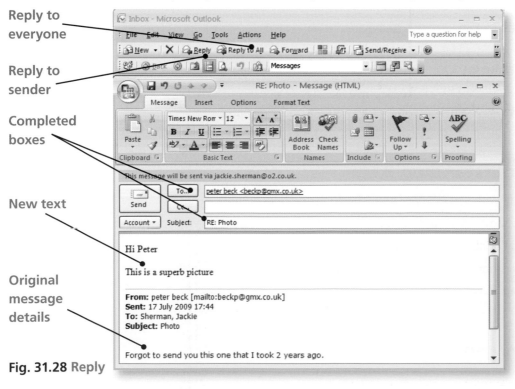

Fig. 31.28 Reply

4 Message replies often retain the original text, to remind senders of the message they sent. You can delete some or all of this text or leave it intact.

5 Click at the top of the message, above the original text and write your reply.

6 Send the message in the normal way.

forward a message

1 Select the message and right click or click on the **Forward** button.
2 Enter the email addresses of new recipients in the **To:** and **Cc:** boxes.
3 The subject text will appear automatically, preceded by **Fw:**.
4 Leave or delete any of the original message text.
5 Enter your own message at the top of the message window.
6 Send the message in the normal way.

Check your understanding 10

1 Open any message you have received.
2 You are going to forward it so click on the appropriate option.
3 Attach a file such as an image or short word processed document.
4 Add the text *I am practising forwarding a message.*
5 Forward it to a new recipient – for example, a different friend or colleague.

Priority

Some messages are particularly important. If you want to make sure a recipient reads and acts on the message urgently, you can attach a **high priority banner**. (There are also low priority banners, but these are unlikely to be used often.)

set priority

1 Write the message.
2 Before sending, click on the **High Importance** button showing a red exclamation mark for a high priority.
3 To set low priority, click on the blue arrow button.
4 Send the message as normal.
5 When received, the red exclamation mark will appear next to the message in the Inbox if **Importance** is one of the field headings displayed.

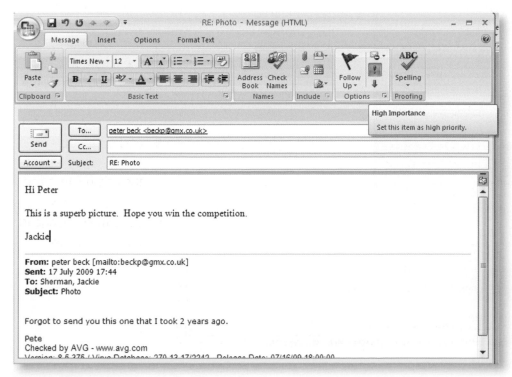

Fig. 31.29 Priority

Flagging messages

A different way to highlight a message is to **flag** it. This is often done when you receive a message that needs action, so that you don't forget it. A coloured flag is placed next to the message and you can change the colour or symbol or remove the flag to indicate the progress of the action taken.

flag a message

1 Right click on the message in the **Inbox** or **Sent items** folder.

2 Select **Follow Up**.

3 Select a suitable time, such as **Today**, or click on a week marker and select the day in the calendar that will appear.

4 Click on **Add Reminder** if you want a reminder such as a sound effect when the date arrives.

5 Once flagged, click on the **flag** to clear it or right click for other options such as to mark it completed.

6 If you click on **Set Quick Click** you can add a flag for your preferred timescale whenever you click on a message.

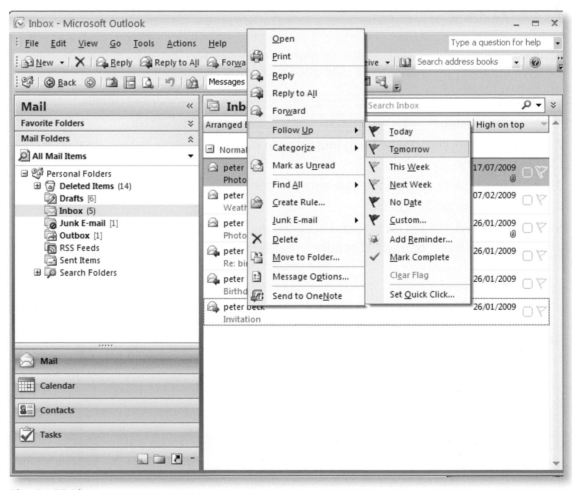

Fig. 31.30 Flag

Check your understanding 11

1 Send yourself a short message with a high priority.

2 When it arrives, flag it to follow up next week.

3 Now clear the flag.

Junk mail

Many messages will be sent to you that are unwanted – usually containing advertisements or more dangerous ones as detailed earlier. These messages are known as **spam** or **junk mail** and can be moved directly to a **Junk Email** folder set up in Outlook. You can then, if necessary, check them before deleting them permanently.

The process is based on rules built into Outlook that identify message senders or key words in the subject of a message. You can use the rules to move messages to different parts of the system automatically. Although Outlook is already set up to identify and move many unwanted messages to the special Junk message folder, you will still receive some that will slip through that you need to deal with.

deal with junk mail

1 When a suspect message arrives, right click on its name in the **Inbox**.

2 If it is a genuine message you want to keep, you can click on the option to add it to a **Safe Senders** list so that it is not removed automatically in future.

3 For unwanted messages, click on **Add Sender to Blocked Senders List**.

4 The current message will be moved to the Junk Email folder and future messages from the same source will be moved there automatically.

5 Click on **Junk Email Options** if you want to remove a sender from the Blocked list or set up more detailed rules for dealing with different types of message.

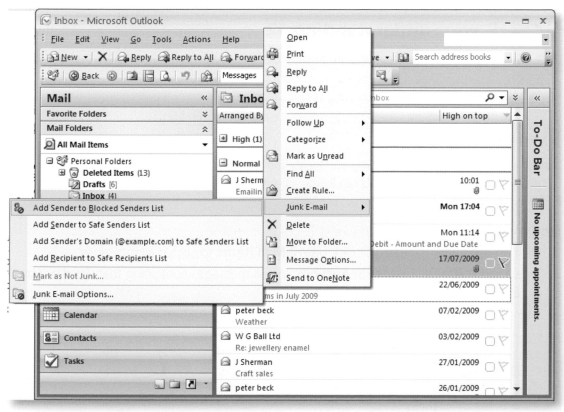

Fig. 31.31 Junk

Sorting messages

As your Inbox or Sent items folders fill up, you need a quick way to organise or find particular messages. Do this by sorting them according to criteria such as date received, who they are from or to, their subject matter and so on.

sort messages

1 Click on one of the field headings, for example **Size** or **From**. The messages will be sorted from largest to smallest or from earliest to latest date received.

2 To reverse the order, click on the same heading again.

OR

3 Right click and select **A–Z Sort Ascending** or **Z–A Sort Descending** to set the order.

4 Add fields if you want to sort by other criteria.

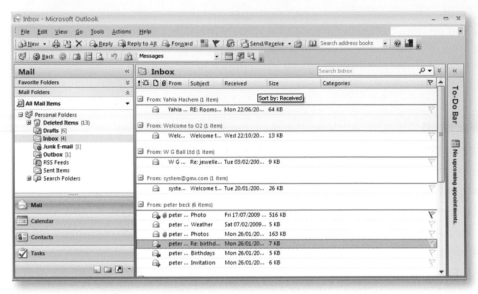

Fig. 31.32 Sort inbox

Check your understanding 12

1 Sort all the messages in your Sent Items folder in alphabetical order of who they are to.

2 Now sort the Inbox in descending order of date received.

Finding messages

As well as setting a sort order to help you locate messages, you can carry out a detailed search by using the **Find** option. Here you can locate all messages based on a named recipient, a word in the **Subject** box or the date sent or received.

find messages

1 Click on the folder to search or click on **All Mail Items** to search all messages.

2 Type a word into the **Instant Search** box and all messages containing that word will be displayed with the word highlighted.

Instant search box

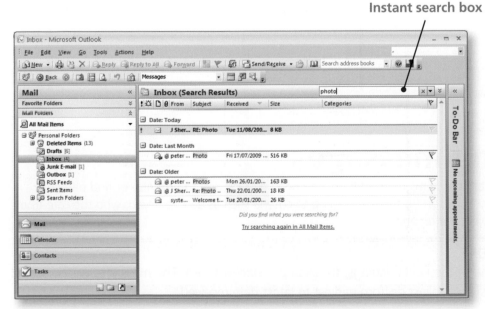

Fig. 31.33
Quick search

3 For a more detailed search, click on the double arrows next to the **search** box to open a new small window.

4 Here you can enter details in the **From**, **To**, **Subject** or **Date** boxes to limit the search, or look for all messages with or without attachments or containing certain text in the main body of the email.

5 Click on the drop-down arrow next to any box heading or the **Add Criteria** button for further criteria such as looking for copied or flagged messages.

Open search window

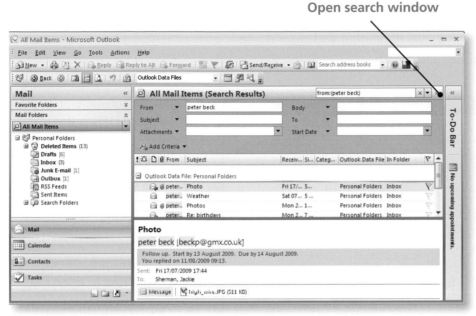

Fig. 31.34 Detailed search

Check your understanding 13

1 Use the search facility to find all messages copied to your own email address.

2 Now search for any files with attachments.

3 Take a screen print to show the results of your search and print a copy (**Print Screen** plus **Paste**).

Fig. 31.35 Find attachment

Printing messages

To make a hard copy of a message, you can print it out. Although emails are meant to herald the 'paperless office', there are many times when you do need a printout –for example, codes for tickets purchased online, or details to take with you relating to meetings or trips.

print an email

1 Right click on and select **Print** to print one copy using default settings.

   Or

2 Go to **File – Print** to open the print dialog box and check in **Print Preview**.

3 You can choose a style of printout and make changes such as adding a header or footer or changing the fonts.

   Or

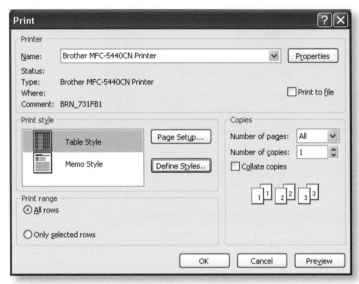

Fig. 31.36 Print message

4 Open the message first and then go to **Office Button – Print**.

5 Click on **Print Preview** to view the message. For example, you may want to make sure message header details are present and any attachments are visible.

6 Click on **Print** to open the **Print** dialog box.

7 Click on **Define Print Styles – Edit** to choose how the printout will appear.

8 You may have the option to print out attachments at the same time.

Check your understanding 14

1 Locate any message.

2 Print out one copy using the default settings.

3 Now make changes to the style or fonts before printing a second copy.

Addresses and contacts

So that you can send email messages, you need to have at hand the email address of anyone you wish to write to. If they have ever sent you an email, their address will be held on your system. But if you see an address in a magazine, on a letter or are given it over the phone, you need to add it to your **Contacts** address book.

Opening Contacts

Contacts is available from the main Outlook window.

open Contacts

1 Click on the folder name in the main window.

2 Depending on the selection, you will see the names arranged as a list or as small cards, with alphabetical tabs down the side to help you move through the addresses.

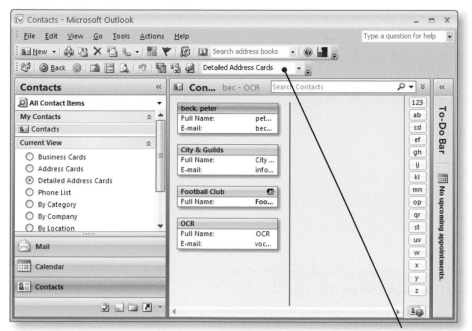

Fig. 31.37 Open contacts

Ways to view contents

Remember

Having someone's personal data is a privilege. Do not share it without that person's permission – you could be giving your friend's details to an identity thief, or someone they've lost touch with for a reason.

add a new contact

1 Click on the **Contacts** folder name.
2 If organised as a phone list, click on the top empty row and type the details here. To add further information, double click on the row and a new Contacts window will open.
   Or
3 Click on the **New** button.
4 In the window that opens, add as much information as required. You must always add the name, full email address and display name.
   Or
5 Select an incoming email message.
6 Right click on the address in the **From:** box.
7 Click on **Add to Outlook contacts**.
8 A New Contact window will open with some of the details already completed.
9 If you want to, add extra details such as postal address, telephone number and so on.
10 Click on the **Save & Close** button to save the details.

Fig. 31.38 Add to contacts

edit a contact

1 Open **Contacts**.
2 Double click on the target name.
3 Change any entries as required.
4 Click on **Save & Close** to update the details.

delete a contact

1 Open **Contacts**.
2 Select the details.
3 Right click and select **Delete**.
   Or
4 Press the **Delete** key.

Check your understanding 15

1 Add the following two contacts to your Contacts folder:
   - Name: John Gray Email address: j.gray@swapshop.co.uk
   - Name: Mary Hopkins Email address: maryhopk@sanderson.net
2 Add the address of your friend taken from one of their messages.
3 Now amend John Gray's address to read j.gray5@swapshop.co.uk.
4 Finally, delete Mary Hopkins' details from your Contacts folder.

Using Contacts when composing

When creating a message, you can add the email address of anyone in Contacts easily.

add Contacts to a message

1 Start the new message.
2 Click on the **To:** button. This opens up a list of your Contacts in a **Select Names** window.

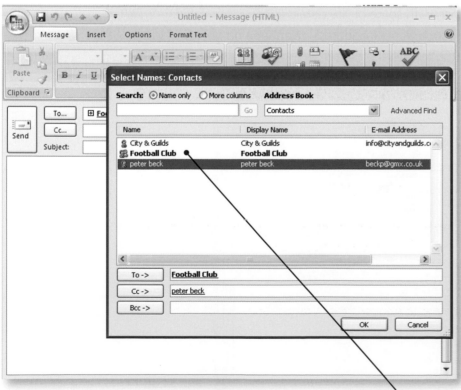

3 Select the recipient in the top window and double click on their details. These will be added to the **To:** box below.
4 Add another recipient in the same way if required.
5 Click on **OK** to return to the message.
6 If sending copies, click on the **Cc:** box in the message and double click on names to add their email addresses in the Cc: box.
7 If you start typing the email address of someone in your Contacts folder you should also be offered their full details automatically.

Fig. 31.39 Add contacts automatically to message   **Grouped addresses**

Group addresses/Distribution lists

If you regularly send the same emails to a number of people, you can group their details together into a distribution list. This will be displayed as a single address in Contacts so that you can click on it to add all the addresses to your message in one go.

create a distribution list

1 Click on the drop-down arrow next to the **New** button and select **Distribution List**.

2 When the window opens, type a name for the group – for example, *Hotel customers.*

3 To add addresses already held in Contacts, click on **Select Members**.

4 Double click on the name of anyone you want to add to the group to place their name in the **Members** box.

5 Continue until everyone has been added and then click on **OK** to close the window.

6 To add further members whose details you do not have in Contacts, click on **Add New**.

7 Enter their details in the boxes and click on **OK**.

8 Click on the checkbox if you also want to add their details to your Contacts folder.

9 Click on **Save & Close** to update the distribution list details.

10 The group title will be visible in Contacts in bold with several faces next to it. Click on its name to add all the addresses to your messages.

Group name   Add other members

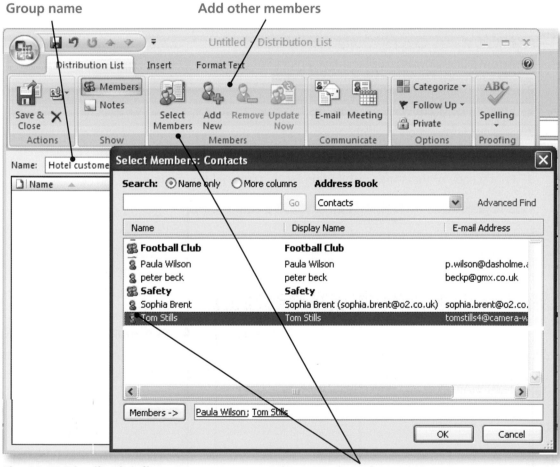

Fig. 31.40 Distribution list   Select members from contacts

Printing Contacts

You can follow normal printing procedures to print out the details of any contact in your Contacts folder.

print a contact

1 Select the Contact name.

2 Click on the **Office** button.

3 Go to **Print – Print Preview** to check how the details will appear.

4 To change any settings such as paper size or orientation, click on the **Page Setup** link.

5 Click on **Print** to produce a copy.

6 Note that an **open** Contact will only print in memo style, but a **closed** Contact can be printed in various styles, depending on what style you are using to view the contents of the folder.

7 (Note also that if printing a distribution list, the group name will appear on page 1 and the members' details on further pages.)

You can also export your contact details for use in other applications. In Outlook's **File** menu, select **Import and Export**, then **Export to a File**. Click **Next**. You can then choose from a range of export options – .csv files readable by a number of software applications, or files specifically for use in programs such as Access or Excel. Select your export option, then click **Next**. Navigate to your **Contacts** folder, and then choose a file name and **Save** your exported contacts.

Check your understanding 16

1 Add the following addresses to your Contacts folder:

   a Mike Houdini: mike_houdini@scarey.net

   b Sophia Brent: sophia.brent@o2.co.uk

   c mail to: wallaceh@virgin.net

2 Create a distribution list named *Magic*.

3 Add Mike and Sophia from your Contacts folder.

4 Add Wallace Howard: wallaceh@virgin.net but do not add him to your Contacts.

5 Save and close the Contacts folder.

6 Reopen the distribution list *Magic*.

7 Remove Sophia from the list.

8 Save this change.

9 Print a copy of the *Magic* distribution list to display all the members.

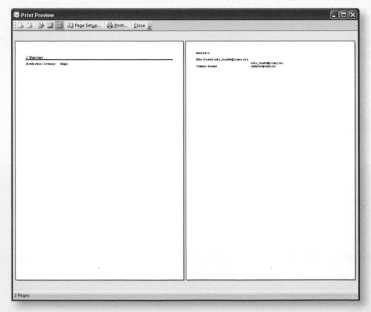

Fig. 31.41 Magic

Organising email

Where you want to work with related messages, appointments or tasks, or need to find them at a glance, Outlook 2007 offers you the chance to highlight them by applying colour coding. You can then sort or search for the same coloured items. This type of organisation is known as colour categories.

apply colour categories

1 Select any item you want to categorise.

2 Click on the **Categorise** button or right click on the item and choose a colour.

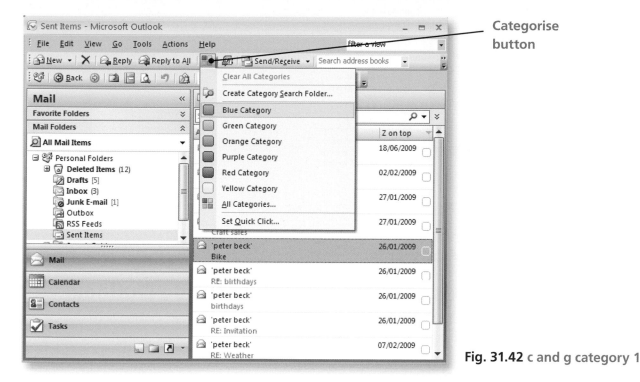

Categorise button

Fig. 31.42 c and g category 1

3 For a new category, you can give it a more meaningful name.

4 Add all relevant items to the category as you create or receive them by clicking the **Categorise** button and selecting the appropriate category.

Fig. 31.43 c and g category 2

sort using categories

1 Open the folder containing categorised items.

2 Drag the Category heading into the space at the top of the window. (If there is no space, click on the **Group By Box** button first.)

3 When you let go, you will now see the messages or contacts grouped by category.

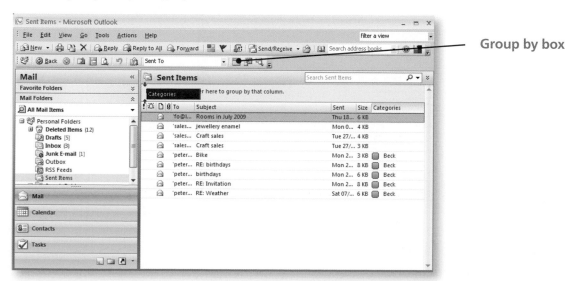

Group by box

Fig. 31.44 c and g sort by category

4 You can also sort by going to **View – Arrange by – Categories**.

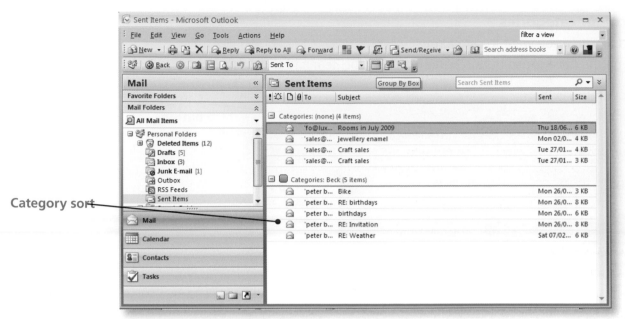

Fig. 31.45 c and g category sort 2

Create and use Office documents inside Outlook

You may create and use Office documents inside Outlook as OLE (Object Linking and Embedding) objects in your emails. This technique allows you to use any of the Office document features, such as spreadsheet formulas in your email.

1 Create a new email.

2 Click on the **Insert** tab at the top of screen.

3 Click on the **Object** button inside the Text section of the **Insert** toolbar.

4 Choose the object type, e.g. **Microsoft Office Excel Worksheet,** then use the **OK** button.

5 Add the text, formulas, formats and whatever else is needed to this OLE document object. You will notice that the toolbars for the document type are there at the top of the screen.

6 You can click away from the OLE document object to continue with the rest of your email.

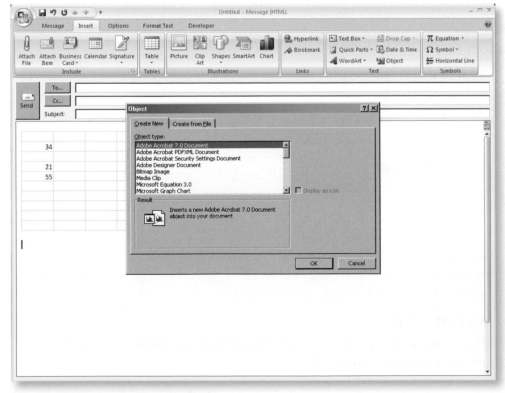

Fig. 31.46 Import OLE to Outlook

Calendar

The calendar section of Outlook is a valuable tool in any office as it allows you to make, schedule, monitor and remind yourself of appointments and meetings. As the different parts of Outlook are linked, it also means that you can invite people to meetings, remind them of venues and so on directly using your built-in Mail and Contacts folders.

There are four different types of item you can add to the Outlook 2007 calendar:
- **Appointments** – where you are interested in the start time, date and venue
- **Events** – these apply to whole days and so you are only interested in the date and venue
- **Meetings** – here you are concerned with start and end times and also who is attending
- **Tasks** – these are actions for you or that can be assigned to different people and may need to be followed up.

open the calendar

1 Click on the **Calendar** folder name in the **Navigation** pane.
2 When it opens, click on a **Day**, **Week** or **Month** button to change the display.
3 Use the back or forward buttons to change the date being displayed or click on the **small calendar** in the **Navigation** pane to display that date in the main window.

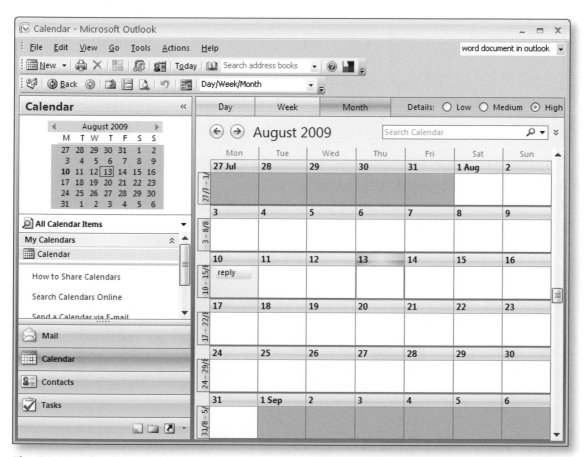

Fig. 31.47 Calendar view

Creating appointments

When you have an appointment, you can add the details to the calendar and use other facilities such as reminders to help you keep all your appointments.

create an appointment

1 Change to **Day** display and type in the appointment details on the right day in the correct time box, to add an appointment quickly.

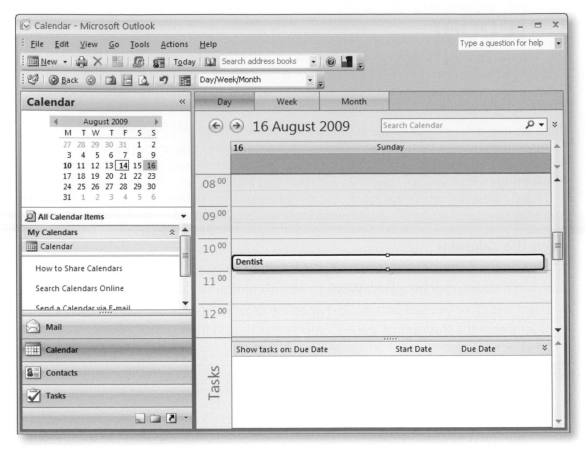

Fig. 31.48 Quick appointment

Or

2 Double click on the date to open the **Appointments** window.

3 Complete the subject and location boxes.

4 Set the start time and, if relevant, set an end time also.

5 Add any special details in the main window.

6 Click on **Save & Close** to save the appointment.

Click for events

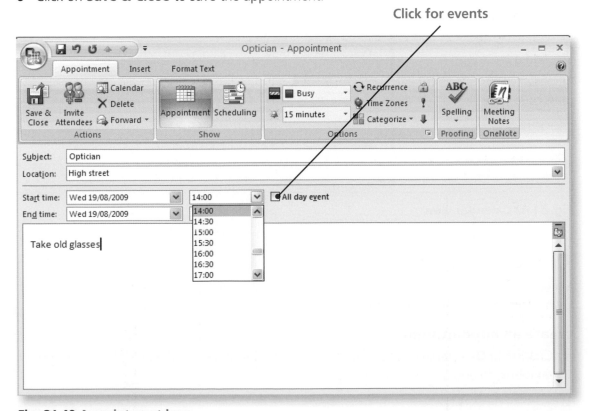

Fig. 31.49 Appointment box

7 The appointment will now be visible in the main window.

8 To view the details, rest the mouse pointer over the text.

9 To reopen the box, double click on the text in the calendar.

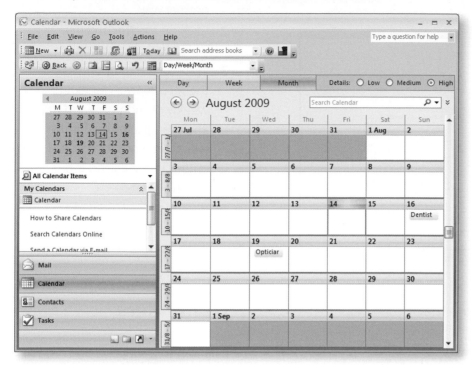

Fig. 31.50 View appointments

delete an appointment

1 Right click on the text in the calendar.

2 Select **Delete**.

reschedule an appointment

1 Locate the appointment in the calendar.

2 Drag and drop it into another time or day slot.

3 You can also open the appointment box and change the details.

set an event

1 Open the appointment window for the correct day.

2 Add a tick in the **All Day Event** box.

3 When saved, an event will be displayed at the top of the day box, rather than at a particular time.

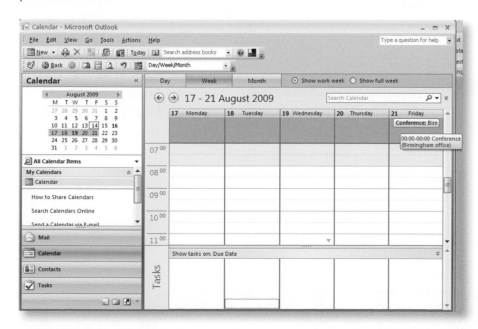

Fig. 31.51 Event

Click for regular event

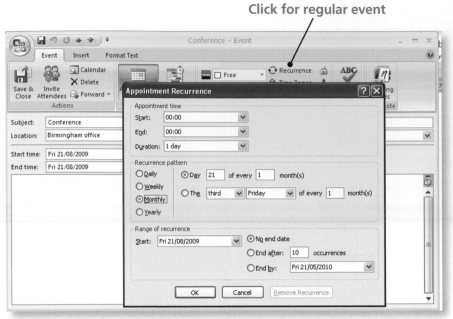

Fig. 31.52 Recurrence

set recurring appointments or events

1 Set the appointment in the normal way.

2 Click on the **Recurrence** command.

3 In the box that opens, set the recurrence – for example, weekly or monthly –and make any changes to days of the week, time or when to end the entries in the calendar before clicking **OK**.

4 Back in the calendar, double curved arrows will be displayed if the event is recurring.

Reminders

To avoid missing an important date in the calendar, you can add reminders in the form of popup windows and even alarm sounds, Outlook sets these at 15 minutes before an appointment and 18 hours before an event by default.

set your own reminder

1 Open the appointment/event window.

2 Click in the **Reminder** box and select your preferred timing.

3 To play a sound, click on the **Sound** button and click for the default sound or browse for your own file.

4 Click on **Save & Close** to save the reminder.

5 When the time arrives, a window will pop up and you can dismiss it or click on **Snooze** for a further reminder at a later time.

Fig. 31.53 Reminder

Fig. 31.54 Reminder popup

Check your understanding 17

1 Open your calendar.

2 Select a day next week and set the following appointment:

   a Meet Hannah for tea

   b Ritz hotel

   c Start time 3.00 p.m.

   d End time 4.00 p.m.

3 Set a reminder for one hour before the appointment.

4 Save and close.

5 Now change the appointment so that the start time is 3.30 p.m.

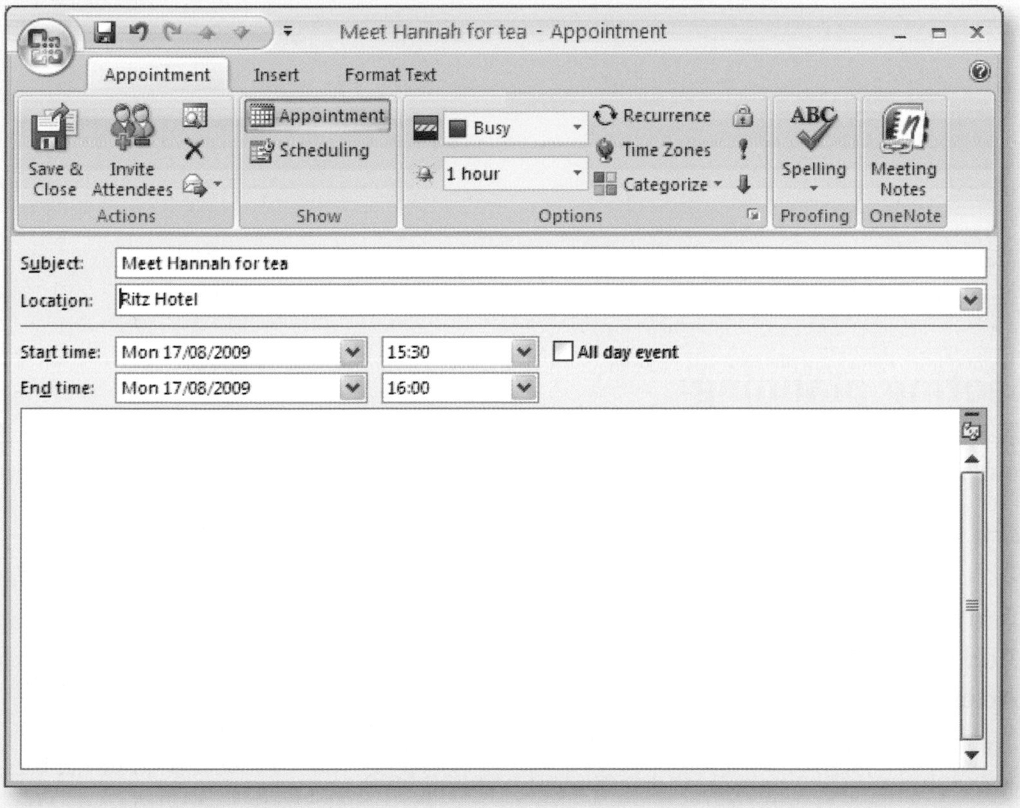

Fig. 31.55 Meet Hannah

Meetings

When setting up meetings, you can invite people to attend and monitor their responses. You can also book resources such as rooms or equipment.

If you are on a network and have access rights, you may be able to access their calendars so that you can check on availability and plan meetings at the best time for everyone.

set up a meeting

1 Go to **New – Meeting Request** or open the appointments window for the time and day of the proposed meeting.

2 Add any relevant notes in the main window.

3 Click on **Invite Attendees** and add everyone's email address. You can add these directly or click on the **To:** button to open your Contacts or global address list folder.

4 Click on **Send** to send the invitations.

5 When the meeting is displayed in the calendar it will include the location and meeting organiser's details.

6 If you cancel a meeting, a cancellation email will be sent to all attendees.

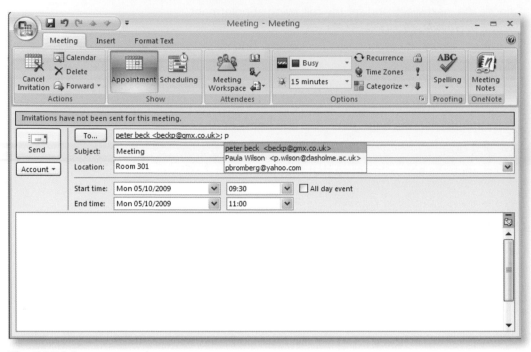

Fig. 31.56 Set up meeting

Meeting planning

For networked systems, colleagues can set their calendars to show periods when they are busy or free. A meeting organiser will then be able to set the meeting time and day for non-busy times.

check schedules

1 Go to **Actions – Plan a meeting** or click on **Scheduling** in the open meeting window.

2 Add the names of people you want to attend by typing in their names or clicking the **Add Others** button and selecting from an address list. Their details will be added to the **Attendees** window.

3 If you have access to their diaries, you will see if they are busy or free of other appointments/meetings at your proposed meeting time.

4 Move the start and/or end times of your proposed meeting by dragging the green and red coloured markers or changing times in the start and end time boxes.

5 When you are happy with the details, click on **Make Meeting** to prepare invitations.

Good time for meeting

Busy periods

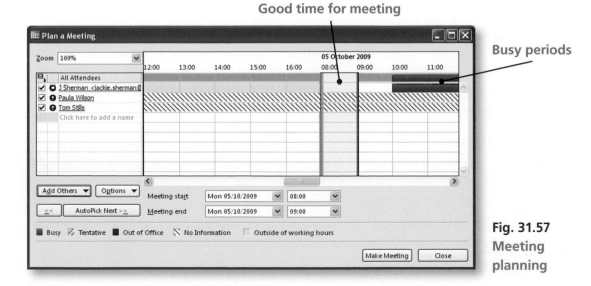

Fig. 31.57 Meeting planning

Receiving a meeting invitation

When an invitation to a meeting arrives, it will display four action buttons: **Accept**, **Decline**, **Tentative** (acceptance) or **Propose New Time**.

respond to a meeting invitation

1 Click on the appropriate button so that the meeting organiser knows whether you can or cannot attend.

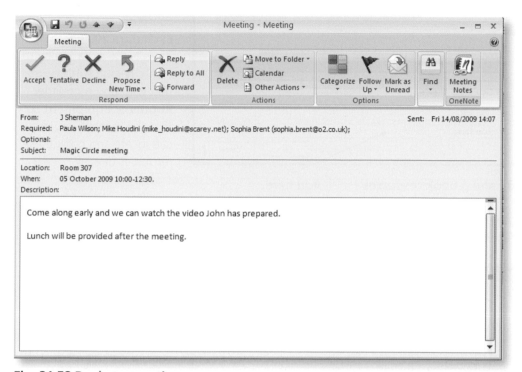

Fig. 31.58 Reply to meeting

2 If you click on **Accept**, a window will open to allow you to add the meeting details to your calendar as well as comments to your automatic reply.

Fig. 31.59 Accept invite

Check your understanding 18

1 Set up a meeting for 2.30–4.30 p.m. next Friday afternoon.

2 Add the names of two people from your Contacts folder who are happy to receive invitations/to reply to messages.

3 If you can, find the details of someone in your organisation whose diary you can check, to see if they would be free. Do NOT invite them (unless they are happy to receive/reply to practice emails).

4 Send an invitation to your two contacts, with the following details:
   • Meeting to discuss new building works
   • In the body of the invitation, ask them to bring their photographs.

5 Send the email invitations.

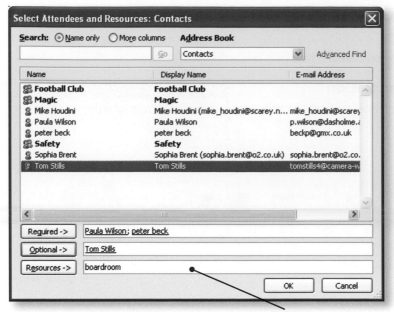

Fig. 31.60 c and g book resources   Add here

Booking meeting resources

When organising a meeting, it can save time if you can also book the room or equipment. For computer users with **Exchange 2007** accounts there is the extra facility to create resource mailboxes that can represent either **Rooms** or **Equipment** so that the meeting organiser can locate and book these resources directly.

When you are booking these items they must be added to the **Resources** category box on the meeting request form.

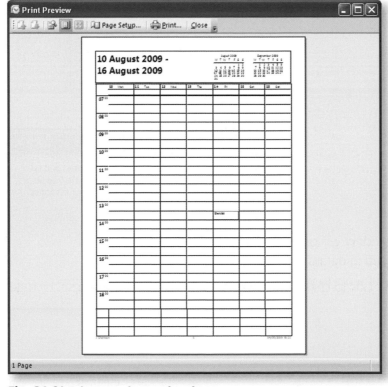

Fig. 31.61 Print preview calendar

Printing the calendar

It is very useful to have a printout of calendars as a reminder of activities such as meetings or courses.

print the calendar

1 Click on the display you want to print – for example by day, week or month.

2 Go to **File – Print Preview** to see how it will look.

3 Click on **Page Setup** to make changes to the display.

4 Click on **Print** to print one or more copies.

Check your understanding 19

1 Select one week in your calendar.

2 Print one copy using the default settings.

Saving a calendar as a web page

It is possible to save Outlook items in different formats, just as you can save other Office files. One useful option is to publish your calendar on the Web so that other people can access it.

save the calendar

1 Go to **File – Save as Web Page**.

2 Under **Duration**, enter start and end dates.

3 Under **Options**, select the options you want such as showing your appointments.

4 Enter a title for the calendar in the **Calendar title** box.

5 In the **File name** box, enter the file protocol and the path to the location where you want to save the file.

6 Take off the tick in the checkbox if you don't want to view it in your browser.

7 Click on **Save** to publish the page.

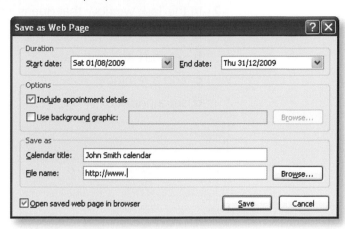

Fig. 31.62 **c and g publish calendar online**

One of the advances in Outlook 2007 is the ability to share and export calendar data, also called 'calendar publishing'.

In calendar view, the management panel includes hyperlinks that allow you to share your calendar, send it via email, or publish your calendar to the Internet. Simply click on the relevant link and follow the on-screen instructions.

When others have shared their calendar with you, remember that you are being given access to their personal information. Do not share calendars without being sure that you have permission to do so.

Tasks

To keep an eye on actions you need to perform you can schedule in tasks, check their progress and set their priority. You can even assign a task to someone else.

Tasks may not need to have exact start or end times, so if they are not completed when the time ends they will be retained until they are marked as completed.

create a task

1 Click on the **Tasks** folder.

2 You will see a new **To-Do** pane showing any current tasks listed.

3 Click on a named task to view the details in the reading pane.

4 Click in the **Current View** box to change the way tasks are listed – for example, by date if overdue, or by category.

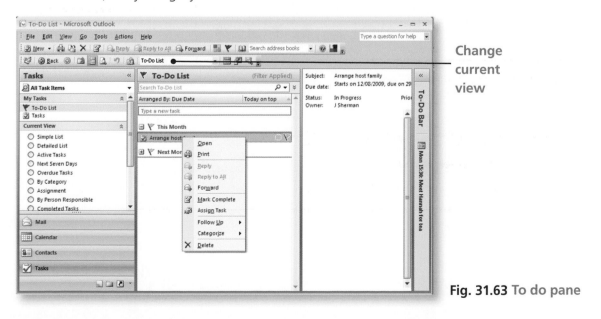

Change current view

Fig. 31.63 **To do pane**

5 To add a task, click on the **Type a new task** box.

Or

6 Click on the **New** button.

7 In the Task window that opens, complete the various boxes to:

  a describe the subject of the task

  b set start and end date

  c add a reminder

  d set a priority.

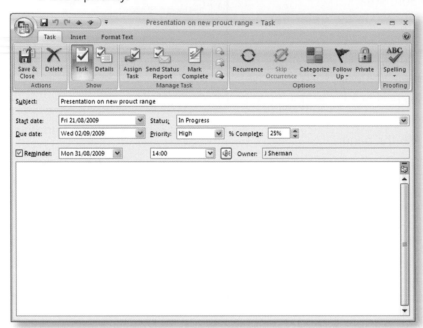

Fig. 31.64 Task

8 Once some time has passed, you can double click on a task in the **To-Do** pane and change the **% Complete** entry to monitor its progress.

9 Right click on **Tasks** in the **To-Do** pane to mark as completed, delete, forward or colour code in a category.

10 If you right click or open the task and click on the **Assign Task** button, you can send the task to another person to complete.

11 To keep track of an assigned task, click on the checkbox to keep an updated copy and ask for notification when the task has been completed.

12 If you are assigned a task, open it and then click on the **Accept** or **Decline** option.

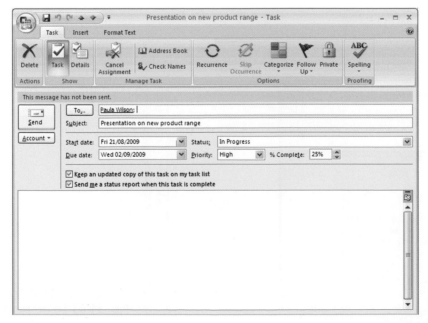

Fig. 31.65 Assign task

Check your understanding 20

1 Create a task. The subject is *Taking photos of the sea.*

2 The end date is a month ahead.

3 So far the task is in progress and 50% has been completed.

4 Set a high priority.

5 Take a screen print showing the task listed in a To-Do list.

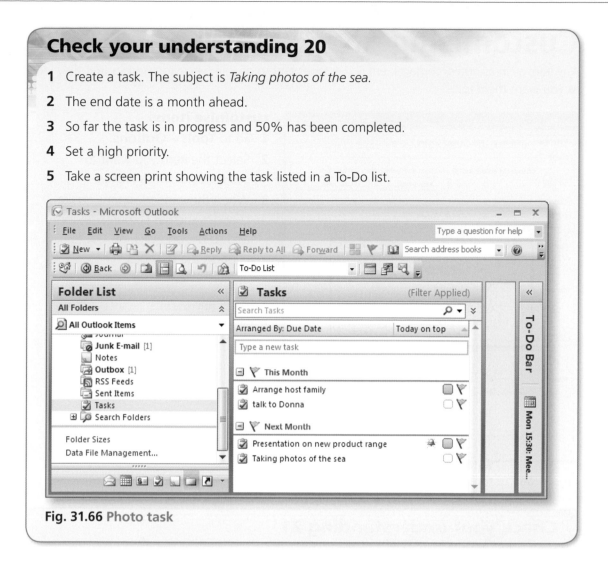

Fig. 31.66 Photo task

Notes

Instead of using post-it notes stuck to the front of your computer, you can create virtual notes as a reminder of short-term actions, addresses and so on that will be readily accessible within Outlook. You could also drag a note from Outlook onto your desktop for immediate accessibility.

create notes

1 Click on the **Notes** icon at the bottom of the navigation pane to view any current Notes.

2 Click on **New** to create a new note.

3 Enter your text and then close the note to add it to the main window.

4 Click on the **Current** view that you prefer, such as displaying notes for the last 7 days or as a list.

5 Change the icon size by selecting an option on the toolbar.

6 To edit a note, double click on it in the main window and retype the text.

7 To remove a note, right click and select **Delete**.

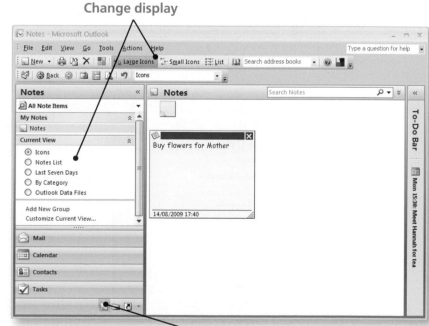

Fig. 31.67 Notes   Notes folder

Customising

Any item such as notes, contacts, email messages and so on can be customised so that they look as you want them to.

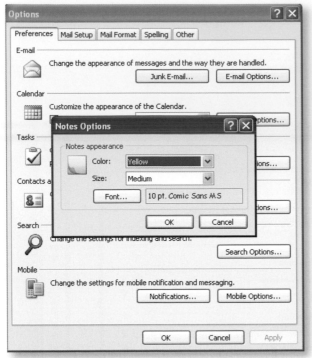

Fig. 31.68 Customise

customise items

1 Go to **Tools – Options**.

2 Select the item you want to customise, such as **Notes**.

3 When the window opens, make changes such as colour, size, fonts and so on, and click on **OK** to confirm the changes.

Check your understanding 21

1 If you can, customise your notes so that they are a different colour.

2 Create a new note.

3 Add the text *Help Harry move house*.

4 Now create a second new note.

5 Add the text *Book tickets for trip to Spain*.

6 Edit the first note so that *Harry* becomes *Jo*.

7 Delete the note about booking tickets.

8 Take a screen print showing your notes.

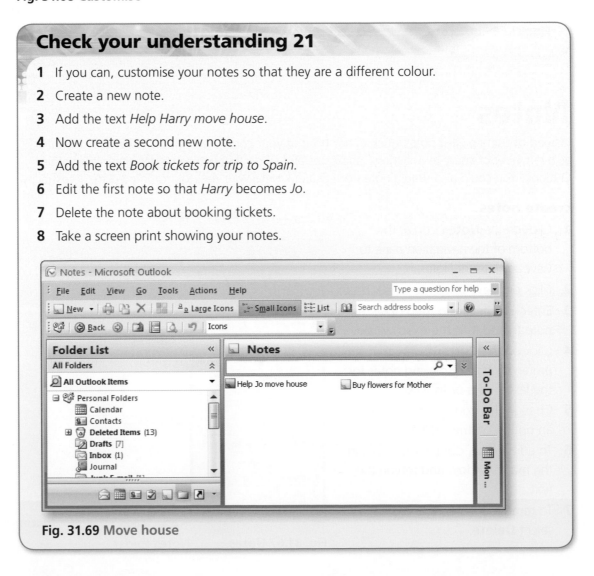

Fig. 31.69 Move house

Integrating different Outlook items

The great value of Outlook is that one item can be turned into another by dragging it to a separate folder. In this way, you can create a task or note from an appointment or mail message, and if a message contains information about special dates, these can be turned into events or appointments by marking them on the calendar.

Note that you can even drag files such as word processed documents or spreadsheets into an Outlook folder on the desktop, if you want to work with them in Outlook.

combine items

1 View the Folders list.

2 Display and select an item such as an appointment in the main window.

3 Drag it with the pointer to a different folder such as Mail, Tasks or Notes.

4 Click on the correct option when a **create** window opens.

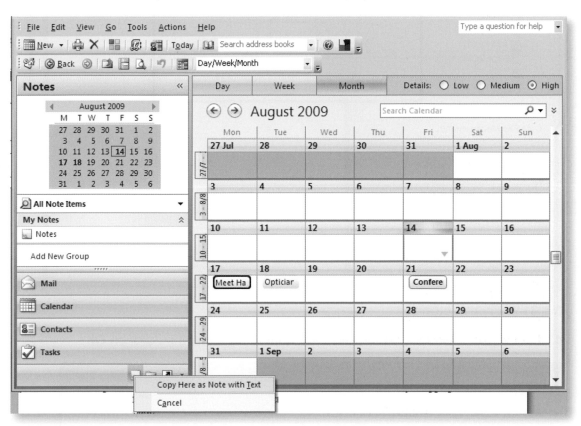

Fig. 31.70 Combining

Journal

One folder that you can use to track your time spent working on particular office documents, sending emails or other activities such as phone calls associated with particular contacts is the **Journal**.

turn on the journal

1 Click on the **Folders list** button to display all the Outlook folders.

2 Click on the **Journal**.

3 Click on **Yes** to turn it on.

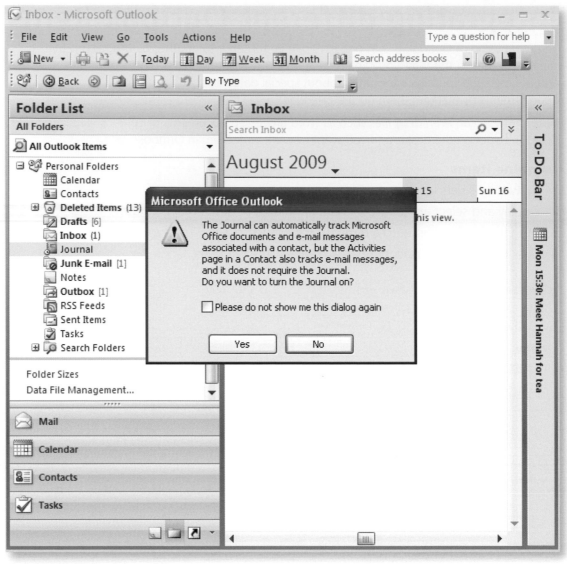

Fig. 31.71 c and g journal

4   In the Options window, set the items you want to track and then click on **OK**.

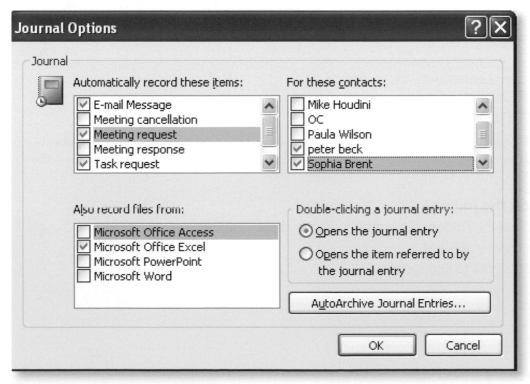

Fig. 31.72 c and g Journal options

5 When the Journal opens, you will see any tracked items on the time line.

6 You can also create journal entries as you work with Outlook.

    a Go to **Actions – Create – New Journal Entry** with the Contacts folder open, to add an action in the Journal such as a phone call to a specific contact.

    b Open the Journal and go to **New – Journal entry** to time work spent on a document or at meetings. After entering the details, click on the **Start Timer** button and then minimise Outlook to carry out the work. When finished, save and close and the time spent will be recorded in the Journal.

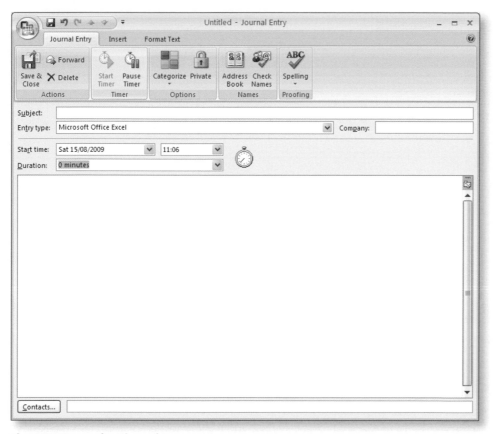

Fig. 31.73 c and g time document

7 To view all activities related to a contact or activity, enter the details in the Journal search box.

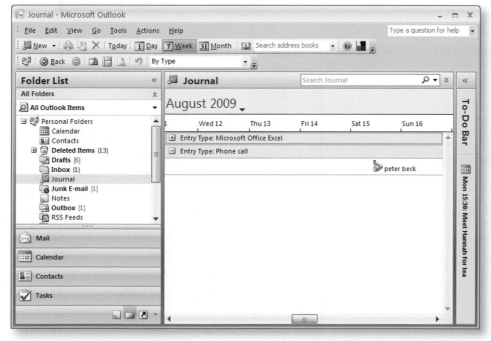

Fig. 31.74 c and g view journal entry

Note that the Journal is not turned on automatically in Outlook 2007 as you can also carry out tasks such as tracking activities with a contact by viewing their details and then clicking on the **Activities** button.

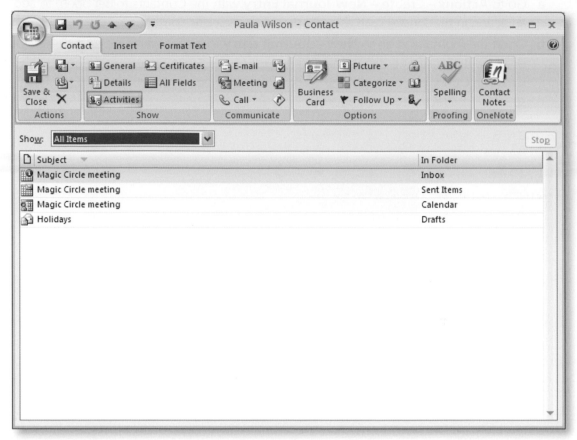

Fig. 31.75 c and g track contact

Managing messages

No one could store all their incoming messages in the single Inbox without quickly becoming overwhelmed by the volume of messages it contains and losing track of related or important messages.

Just as with your computer files, you need to create folders in which to store messages and also to delete any unwanted messages as often as possible. Where messages need to be kept for a certain period of time, you can use the archive facility to move the messages to a storage folder which will then free up space in your Inbox.

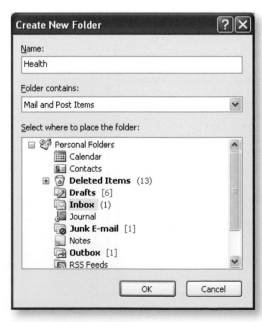

Fig. 31.76 New folder

create new folders

1 Click to select the parent folder such as the **Inbox**.

2 Right click or click on the down arrow next to the **New** button and select **New Folder**.

3 In the window that opens, type in a name for the folder and, if necessary, change the parent folder by clicking a different folder in the **Select** box.

4 Click on **OK** and the new folder will appear below the parent folder.

5 Repeat this process to create more folders or subfolders for any mail folder.

6 To rename a folder, right click and select this option.

7 To delete an unwanted folder (after first moving out any messages you want to keep), select it and then press the **Delete** key. Check before clicking on **OK** when the warning appears.

Fig. 31.77 Delete folder

Moving messages

Once you have set up a folder structure, you can move messages into appropriate folders – or add copies – in different ways:
- physically by dragging
- using the **Move Items** window to select the destination folder
- setting up rules to move messages automatically.

move messages by dragging

1 Open the mail folder containing the messages you want to move.

2 Select one or more messages. Hold down **Ctrl** to select more than one if they are not adjacent.

3 Drag these across to the folders list using the mouse.

4 To copy a message, hold down **Ctrl** as you drag.

5 When the destination folder turns blue, let go of the mouse and the messages will drop inside.

6 If you drag with the right mouse button, a menu will appear when you let go and you can select a **move** or **copy** option or cancel the action.

Destination folder

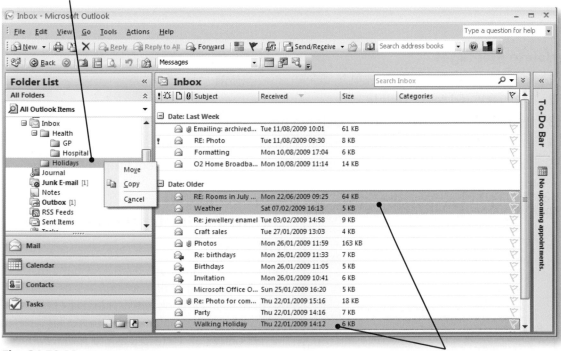

Fig. 31.78 Move message

Selected messages

move messages using Move Items

1 Select the messages you want to move.

2 Right click and select **Move to Folder**.

3 In the **Move Items** window, select the destination folder. Click on a + sign to select a subfolder hidden inside.

4 Click on **OK** and the messages will be moved.

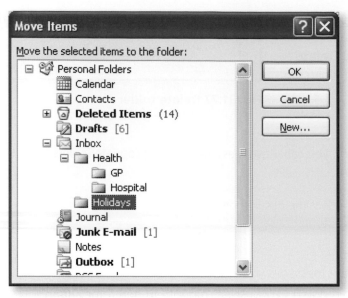

Fig. 31.79 Move items

Check your understanding 22

1 Create two new Inbox folders: label one *Time* and the other *Space*.

2 Select two non-adjacent messages in your Inbox and move them into the *Time* folder.

3 Select one message in the Inbox and copy it to the *Space* folder.

4 Now use any method to move the two messages back out of the *Time* folder into the Inbox.

5 Delete both the *Time* and *Space* folders.

delete messages

1 Select the message(s) you want to delete.

2 Right click and select **Delete**.

3 The messages are moved to the **Deleted Items** folder.

4 To restore a message deleted by mistake, simply use a move option to move it out again.

5 To delete messages permanently, right click on the folder and select **Empty Deleted Items folder**.

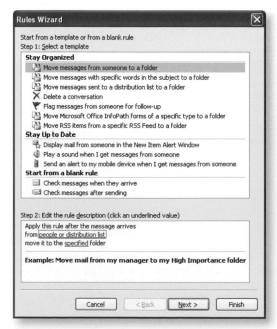

**Fig. 31.80
Rules wizard**

Using rules to move messages

Earlier in the chapter you learned about junk mail. The removal of unwanted messages makes use of rules in Outlook where messages from certain originators or containing particular words are recognised and directed to the **Junk Mail** folder.

You can create your own rules to recognise particular messages such as those from a named source and move them to designated folders you have created. New messages will still show up when they arrive as a number will appear next to the folder name showing it contains unread messages.

set up a rule

1 Click on **Tools – Rules and Alerts**.

2 Click on **New Rule**.

3 Select the appropriate rule from the list – for example, to move messages from someone to a specified folder.

4 In the next window, click on the underlined text in the **Step 2: Edit** box to select the named author of the messages and the destination folder.

5 There are opportunities to add extra conditions or exceptions and to name the rule before it is applied.

6 At any time, you can return to the **Rules and Alerts** window to change or delete the rule.

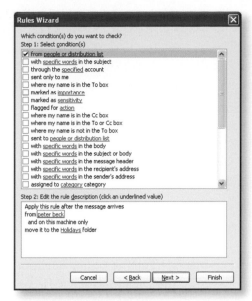

Fig. 31.81 Rules wizard2

Archiving

Archived messages are stored in an archive folder and so can be accessed when necessary.

You can archive an important message manually or make use of the **AutoArchive** facility. This takes a regular sweep of Outlook and identifies messages older than the date you have set. The messages will be moved automatically and, if you choose, can also be deleted after they reach a particular age.

archive a message

1 Select the message. The whole folder in which it is stored will have the same archive features set.

2 Go to **File – Archive**.

3 In the window that opens, set the **older than** date for when messages will be moved.

4 Click on **OK** and a new Archive folder will be created and will be visible in your Folders list.

Fig. 31.82 Archive

use AutoArchive

1 Go to **Tools – Options**, click on the **Other** tab and select **AutoArchive**.

2 You can now set the timing for a sweep of the system, the dates for moving or deleting old messages and the file location in which to store the messages.

3 If you set the same timings for all folders, you can still change individual settings for a folder by right clicking, selecting **Properties – AutoArchive** and taking off the settings.

4 If you set a prompt, you will be asked regularly if you want to cancel the archiving or let it run.

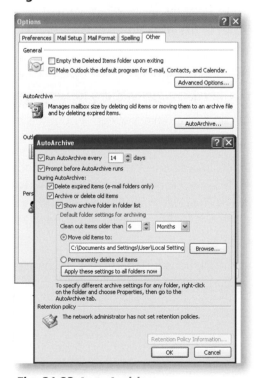

Fig. 31.83 AutoArchive

Assignment

This practice assignment is made up of four tasks

- Task A - Navigation and facilities
- Task B – Set up contacts and distribution lists
- Task C – Send and organise emails
- Task D - Use the calendar and set tasks

You will need the following files:

- Advert.jpg

Scenario

You work as a bookings manager for The Pickled Beetroot, a local public house. In this role you need to be able to organise the entertainments schedule of this venue for bands and other entertainment.

You have been asked to use Outlook to send emails, organise events, set up contacts for bands and keep a calendar.

The publican also wants a simple guide on how you've set up Outlook.

Please read the text carefully and complete the tasks in the order given.

Task A - Navigation and facilities

1 Open Microsoft Outlook and take a screen shot. Annotate the screen shot to show the **Mail**, **Calendar**, **Contacts** and **Tasks** components and how to navigate between them.

2 Use **Help** to show how to customise the menu by adding keyboard accessible commands.

   Print the Help page.

3 Create folders for bands and the press in the **Mail** component.

   Add subfolders under the bands folder for these groups:

   Brain Drain
   Demon Noise
   Dire Noise
   The Combine Harvesters

   Take a screen shot showing the folders.

4 Take a screen shot of the **Junk Email** options and annotate it to explain how they can be used to filter out unwanted emails.

5 Sort your emails into date order with the oldest first in the view. Take a screen shot of this.

   Sort your emails into date order with the newest first in the view. Take a screen shot of this.

6 Create an email signature. Take a screen shot of this being set.

7 Create some new tasks into your **To Do** list:

   a Contact Brain Drain to check availability for Monday week due this Friday
   b Cancel Dire Noise booking for next Wednesday week, due by next Monday
   c Place advert in *The Daily Post* for next week's gigs, due this Thursday
   d Book The Demons for Thurs week, due by next Monday

   Add reminders to these.

   Mark booking The Demons for Thurs week as completed.

   Print out the tasks as a detailed list.

8   Add categories named:

  a   Bookings
  b   Press
  c   Bands

Assign items from task 7a, b, d to the Bookings category.

Assign item from task 7c to the Press category.

Take a screen shot of the sorted categories.

Task B – Set up contacts and distribution lists

1   If you do not have anyone to reply to your emails, set up email accounts using Google or another provider to represent the bands:

Brain Drain
Demon Noise
Dire Noise
The Combine Harvesters

Create records for the above bands in Outlook contacts. Categorise as Bands.

Create a record for *The Daily Post* in Outlook contacts.

Take a screen shot of the contacts organised by category

2   Sort your contacts into full name order, A-Z. Take a screen shot of this.

Sort your contacts into full name order, Z-A. Take a screen shot of this.

Task C – Send and organise emails

1   Read an email using Outlook. Print the email in memo style.

2   Reply to an email.

Print the reply in memo style.

3   Forward an email to Brain Drain. Take a screen shot of this in the sent folder.

4   Send an email to Dire Noise cancelling their booking for next Wednesday week.

Type the email address (do not select from Contacts).

Add the signature you created in task A5.

Flag it as high importance, and request a read receipt.

Take screen shots of these features and options being used.

5   Send emails to book The Demons for next Wednesday week and Brain Drain for next Monday week.

Select the email addresses from Contacts.

Add the signature you created in task A5.

Print the emails in memo style.

6   Send an email to place an advert in *The Daily Post* for next week's gigs.

Select the email address from Contacts.

Add the signature you created in task A5.

Use a style of your choice for this email.

Insert the file, **Advert.jpg**, as an attachment.

Take screen shots of the style being selected and the file being inserted.

7   Find all the email messages sent to Brain Drain.

Take a screen shot of the found emails.

8   Move emails from the sent folder to their appropriate band folders.

Take a screen shot of all the emails in band subfolders.

9   Use auto-archive on your email messages.

Take a screen shot of auto-archive being selected.

10 Select and delete all the emails in your Junk Email folder.

Take a screen shot of the empty Junk Email folder.

Task D – Use the calendar and set tasks

1   Schedule these event appointments into the calendar:

   a) Brain Drain for Monday week, 18:30-23:30

   b) The Demons for Thursday week, 18:30-23:30

2   Schedule a 48-hour "Best of the '60s" multiday event starting Friday week at 18:30 and finishing Sunday week at 18:30.

3   Schedule a recurring Karaoke event for Wednesday evenings 20:00-22:30.

4   Schedule a meeting with the bands for Tuesday week at 15:00-16:00, to discuss a new sound system:

Brain Drain
Demon Noise
Dire Noise
The Combine Harvesters

5   Print month and week views of the calendar.

6   Create a note that says "SM58, Beta87" to remind you of microphone types to be discussed at the band meeting.

Categorise the note as *Bands*.

Take a screenshot of the notes folder.

7   Edit the note to add "SM57".

Take a screen shot of the notes folder.

Appendix

Achieving Level 3

The bulk of information needed to achieve the ITQ qualification at Level 3 has been integrated within the text and assignments found in the preceding pages. However, some of the learning objectives have not been covered in detail. A list of those objectives can be found below, divided by unit.

If you are aiming for the Level 3 qualification, you can stretch yourself by including references to the objectives listed below in your assignments. A Level 3 student will need to become adept at self-management and research, and stretching yourself is an excellent way to achieve this.

Optimise IT System Performance Level 3

1. **Keep computer hardware and software operating efficiently**

    1.1 Explain the factors that should be taken into account when choosing an operating system
    1.3 Explain why routine fault-finding procedures are important
    1.4 Use an appropriate fault-finding procedure to routinely monitor hardware performance
    1.7 Configure synchronisation and maintain security on remote access sessions

2. **Manage files to maintain and improve performance**

    2.1 Explain why it is important to undertake file housekeeping of the information stored on computer systems and how it affects performance
    2.3 Archive, back up and restore files and folders
    2.5 Configure access to remote file systems

3. **Troubleshoot and respond to IT system problems quickly and effectively**

    3.1 Assess IT system problems, explain what causes them and how to respond to them and avoid similar problems in future
    3.2 Carry out contingency planning to recover from system failure and data loss
    3.3 Monitor and record IT system problems to enable effective response
    3.4 Monitor system settings and adjust when necessary
    3.5 Explain when and where to get expert advice
    3.6 Help others to select and use appropriate resources to respond to IT system problems

4. **Plan and monitor the routine and non-routine maintenance of hardware and software.**

    4.1 Clarify the resources that will be needed to carry out maintenance
    4.2 Develop a plan for the maintenance of IT hardware and software
    4.3 Monitor the implementation of maintenance plans, updating them where necessary

5. **Review and modify hardware and software to maintain performance**

    5.1 Use appropriate techniques to maintain software for optimum performance
    5.2 Clarify when and how to upgrade software
    5.3 Review and modify hardware settings to maintain performance

Word Processing Software Level 3

1. **Enter and combine text and other information accurately within word processing documents**

   1.1 Summarise what types of information are needed for the document and how they should be linked or integrated

   1.3 Create, use and modify appropriate templates for the different types of documents

   1.4 Explain how to combine and merge information from other software or multiple documents

   1.7 Select and use tools and techniques to work with multiple documents or users

   1.8 Customise interface to meet needs

2. **Create and modify appropriate layouts, structures and styles for word processing documents**

   2.1 Analyse and explain the requirements for structure and style

   2.3 Define and modify styles for document elements

   2.4 Select and use tools and techniques to organise and structure long documents

3. **Use word processing software tools and techniques to format and present documents effectively to meet requirements**

   3.1 Explain how the information should be formatted to aid meaning

   3.3 Select and use appropriate page and section layouts to present and print multi-page and multi-section documents

   3.5 Evaluate the quality of the documents produced to ensure they are fit for purpose

   3.6 Respond appropriately to any quality problems within documents to ensure that outcomes meet needs and are fit for purpose

Using the Internet Level 3

1. **Select and set up an appropriate connection to access the Internet**

   1.2 Explain the benefits and drawbacks of different connection methods

   1.3 Analyse the issues affecting different groups of users

   1.4 Select and set up an Internet connection using an appropriate combination of hardware and software

   1.5 Recommend a connection method for Internet access to meet identified needs

   1.6 Diagnose and solve Internet connection problems

2. **Set up and use browser software to navigate web pages**

   2.2 Explain when to change browser settings to aid navigation

   2.3 Adjust and monitor browser settings to maintain and improve performance

   2.4 Explain when and how to improve browser performance

   2.5 Customise browser software to make it easier to use

3. **Use browser tools to search effectively and efficiently for information from the Internet**

   3.2 Evaluate how well information meets requirements

4. **Use browser software to communicate information online**

   4.3 Share and submit information online using appropriate language and moderate content from others

5. **Develop and apply appropriate safety and security practices and procedures when working online**

   5.1 Explain threats to system performance when working online
   5.3 Explain the threats to information security and integrity when working online
   5.4 Keep information secure and manage user access to online sources securely
   5.5 Explain the threats to user safety when working online
   5.7 Develop and promote laws, guidelines and procedures for safe and secure use of the Internet

Presentation Software Level 3

1. **Input and combine text and other information within presentation slides**

   1.1 Explain what types of information are required for the presentation

2. **Use presentation software tools to structure, edit and format presentations**

   2.1 Explain when and how to use and change slide structure and themes to enhance presentations
   2.2 Create, amend and use appropriate templates and themes for slides
   2.3 Explain how interactive and presentation effects can be used to aid meaning or impact
   2.5 Create and use interactive elements to enhance presentations
   2.6 Select and use animation and transition techniques appropriately to enhance presentations

3. **Prepare interactive slideshow for presentation**

   3.1 Explain how to present slides to communicate effectively for different contexts
   3.2 Prepare interactive slideshow and associated products for presentation
   3.4 Evaluate presentations, identify any quality problems and discuss how to respond to them
   3.5 Respond appropriately to quality problems to ensure that presentations meet needs and are fit for purpose

Spreadsheet Software Level 3

2. **Select and use appropriate formulas and data analysis tools and techniques to meet requirements**

   2.1 Explain what methods can be used to summarise, analyse and interpret spreadsheet data and when to use them
   2.4 Select and use forecasting tools and techniques

3. **Use tools and techniques to present, format and publish spreadsheet information**

   3.1 Explain how to present and format spreadsheet information effectively to meet needs
   3.3 Select and use appropriate tools and techniques to generate, develop and format charts and graphs
   3.5 Explain how to find and sort out any errors in formulas
   3.6 Check spreadsheet information meets needs, using IT tools and making corrections as necessary
   3.7 Use auditing tools to identify and respond appropriately to any problems with spreadsheets

Database Software Level 3

1. **Plan, create and modify relational database tables to meet requirements**

    1.1 Explain how a relational database enables data to be organised and queried

    1.2 Plan and create multiple tables for data entry with appropriate fields and properties

    1.3 Set up and modify relationships between database tables

    1.4 Explain why and how to maintain data integrity

2. **Enter, edit and organise structured information on a database**

    2.1 Design and create forms to access, enter, edit and organise data in a database

3. **Use database software tools to create, edit and run data queries and produce reports**

    3.1 Explain how to select, generate and output information from queries according to requirements

Index